ENERGY SCIENCE, ENGINEERING AND TECHNOLOGY

SOLAR PHOTOVOLTAICS

COST AND ECONOMIC PERFORMANCE CONSIDERATIONS

ENERGY SCIENCE, ENGINEERING AND TECHNOLOGY

Additional books in this series can be found on Nova's website under the Series tab.

Additional E-books in this series can be found on Nova's website under the E-book tab.

ENERGY POLICIES, POLITICS AND PRICES

Additional books in this series can be found on Nova's website under the Series tab.

Additional E-books in this series can be found on Nova's website under the E-book tab.

ENERGY SCIENCE, ENGINEERING AND TECHNOLOGY

SOLAR PHOTOVOLTAICS

COST AND ECONOMIC PERFORMANCE CONSIDERATIONS

GRIFFIN MARKS
AND
CLAUDIA J. UNGER
EDITORS

New York

Copyright © 2013 by Nova Science Publishers, Inc.

All rights reserved. No part of this book may be reproduced, stored in a retrieval system or transmitted in any form or by any means: electronic, electrostatic, magnetic, tape, mechanical photocopying, recording or otherwise without the written permission of the Publisher.

For permission to use material from this book please contact us:
Telephone 631-231-7269; Fax 631-231-8175
Web Site: http://www.novapublishers.com

NOTICE TO THE READER

The Publisher has taken reasonable care in the preparation of this book, but makes no expressed or implied warranty of any kind and assumes no responsibility for any errors or omissions. No liability is assumed for incidental or consequential damages in connection with or arising out of information contained in this book. The Publisher shall not be liable for any special, consequential, or exemplary damages resulting, in whole or in part, from the readers' use of, or reliance upon, this material. Any parts of this book based on government reports are so indicated and copyright is claimed for those parts to the extent applicable to compilations of such works.

Independent verification should be sought for any data, advice or recommendations contained in this book. In addition, no responsibility is assumed by the publisher for any injury and/or damage to persons or property arising from any methods, products, instructions, ideas or otherwise contained in this publication.

This publication is designed to provide accurate and authoritative information with regard to the subject matter covered herein. It is sold with the clear understanding that the Publisher is not engaged in rendering legal or any other professional services. If legal or any other expert assistance is required, the services of a competent person should be sought. FROM A DECLARATION OF PARTICIPANTS JOINTLY ADOPTED BY A COMMITTEE OF THE AMERICAN BAR ASSOCIATION AND A COMMITTEE OF PUBLISHERS.

Additional color graphics may be available in the e-book version of this book.

Library of Congress Cataloging-in-Publication Data

ISBN: 978-1-62257-651-7

Published by Nova Science Publishers, Inc. † New York

CONTENTS

Preface vii

Chapter 1 The Impact of Different Economic Performance Metrics on the Perceived Value of Solar Photovoltaics 1
National Renewable Energy Laboratory

Chapter 2 Building-Integrated Photovoltaics (BIPV) in the Residential Sector: An Analysis of Installed Rooftop System Prices 51
National Renewable Energy Laboratory

Chapter 3 Insuring Solar Photovoltaics: Challenges and Possible Solutions 99
National Renewable Energy Laboratory

Index 147

PREFACE

Photovoltaic (PV) systems are installed by several types of market participants, ranging from residential customers to large-scale project developers and utilities. Each type of market participant frequently uses a different economic performance metric to characterize PV value because they are looking for different types of returns from a PV investment. Choice of economic performance metric by different customer types can significantly shape each customer's perception of PV investment value and ultimately their adoption decision. In this book, PV economic performance is characterized for three ownership types: residential customers who purchase their own PV systems, commercial customers (for-profit companies) who purchase their own PV systems, and residential and commercial customers who lease PV equipment or buy PV electricity from a third-party company.

Chapter 1 – Photovoltaic (PV) systems are installed by several types of market participants, ranging from residential customers to large-scale project developers and utilities. Each type of market participant frequently uses a different economic performance metric to characterize PV value because they are looking for different types of returns from a PV investment. The authors find that different economic performance metrics frequently show different price thresholds for when a PV investment becomes profitable or attractive. Additionally, several project parameters, such as financing terms, can have a significant impact on some metrics [e.g., internal rate of return (IRR), net present value (NPV), and benefit-to-cost (B/C) ratio] while having a minimal impact on other metrics (e.g., simple payback time). As such, the authors find that the choice of economic performance metric by different customer types can significantly shape each customer's perception of PV investment value and ultimately their adoption decision.

In this analysis, we characterize PV economic performance for three ownership types: residential customers who purchase their own PV systems, commercial customers (for-profit companies) who purchase their own PV systems, and residential and commercial customers who lease PV equipment or buy PV electricity from a third-party company. We characterize the differences in PV economics for each customer based on the different tax implications of ownership. We do not characterize PV economics for large-scale PV developers or utilities because they frequently use complex project financing structures that are beyond the scope of this analysis.

We compare PV economic returns for different PV customers using the following economic performance metrics:

- Net present value (NPV)
- Profitability index (PI)
- Benefit-to-cost (B/C) ratio
- Internal rate of return (IRR)
- Modified internal rate of return (MIRR)
- Simple payback and time-to-net-positive-cash-flow (TNP) payback
- Annualized monthly bill savings (MBS)
- Levelized cost of energy (LCOE).

We characterize relative PV economics for each metric over a range of system characteristics, including PV system price and several non-price parameters including financing terms, tax rates, electricity rates and assumed rate escalations, and PV system performance. Key findings include:

- Different economic performance metrics can show unique price thresholds for when a PV investment becomes profitable or attractive.
- In some cases, the choice of an economic performance metric could have as much impact on the representation of value as decreasing (or increasing) PV prices by up to a factor of four.
- Varying non-price system characteristics, such as financing terms or assumed electricity rate increases, can impact PV economic performance as much as decreasing (or increasing) PV prices by several dollars per watt.
- At higher PV prices, commercial projects may generate higher returns than residential projects because commercial customers can depreciate the capital invested in a PV project. At lower PV prices, commercial projects may generate lower returns because the gain from capital

depreciation is offset by the loss from valuing PV based on tax-deductible energy costs.
- IRR is a poor metric for characterizing the value of U.S. PV systems because the upfront nature of financed PV costs and incentives can lead to an inflated perception of value.
- MIRR and simple payback times show very little sensitivity varying several project parameters, and customers using these metrics may be less likely to be incentivized by policy measures targeting non-price system parameters.
- The MBS metric may generate attractive returns at higher PV prices than other metrics and may be effective at stimulating PV markets. However, the third-party PV companies that frequently market systems based on MBS have different tax structures and costs of capital, and it is unclear whether this will lead to higher or lower relative returns.
- The upfront nature of U.S. PV incentives (e.g., federal investment tax credit and accelerated capital depreciation for commercial customers) can lead to very different PV returns for U.S. systems relative to identical systems located in different countries that are described in the international PV literature.

The different price thresholds for when a PV investment becomes profitable or attractive and the different sensitivities to varying system parameters have significant implications for policy design. For example, if policy is introduced to improve PV financing terms, it could preferentially stimulate market segments where customers use metrics that are sensitive to financing terms (IRR, NPV, and B/C ratios) while having little or no impact on market segments where customers use metrics that are insensitive to financing terms (simple payback). There is also a strong potential for stimulating U.S. PV demand using new mechanisms in addition to traditional incentives focused on reducing PV prices or increasing revenues. These range from simply educating potential customers about the value of a PV investment as seen through different economic performance metrics, to providing access to long-term low-cost financing, to allowing third-party companies to develop simple PV products that can generate MBS.

Chapter 2 – For more than 30 years, there have been strong efforts to accelerate the deployment of solar-electric systems by developing photovoltaic (PV) products that are fully integrated with building materials. Despite these efforts and high stakeholder interest in building-integrated PV (BIPV), the deployment of PV systems that are partially or fully integrated with building materials is low compared with rack-mounted PV systems, accounting for about 1% of the installed capacity of distributed PV systems worldwide by the end of 2009. In this report, we examine the cost drivers and performance considerations related to BIPV for residential rooftops. We also briefly review the history of BIPV product development and examine market dynamics that have affected commercialization and deployment.

As with many renewable energy technologies, system prices—in terms of dollars per installed watt of direct-current peak power capacity ($/W$_p$ DC)—have a significant effect on PV deployment. In general, the installed prices of BIPV systems are higher than PV system prices, but the cause of these price premiums—higher costs, higher margins, or other considerations— and the potential for price reductions remain uncertain. Using a bottom-up analysis of components and installation labor costs, we explore the cost trade-offs that affect the prices of residential rooftop BIPV systems. We compare the prices of three hypothetical BIPV systems with the price of a rack-mounted crystalline silicon (c-Si) PV system, the "PV Reference Case," which is the most commonly installed residential system technology. One of the BIPV cases is a derivative of the c-Si PV case ("BIPV Derivative Case"), and the other two BIPV cases are based on an analysis of thin-film technologies (Table ES-1). In today's solar market, few BIPV products are fully integrated with building materials as envisioned in these BIPV cases; therefore, the cases should be seen as near-term possibilities. In contrast, the PV Reference Case represents a 2010 benchmark system price from an NREL study that uses the same methodology to assess objective system prices. Comparing the hypothetical near-term BIPV cases with the 2010 PV benchmark does not account for the continued advancements and cost reductions in rack-mounted PV systems. Thus, the potential cost advantages we have identified for BIPV installations are likely to change. Additionally, the authors' analysis assumes that economies of scale and installer experience are equivalent for the PV Reference Case and the BIPV cases.

Chapter 3 – Although the market for insurance products that cover photovoltaic (PV) systems is evolving rapidly, PV developers in the United States are concerned about the cost and availability of insurance. Annual insurance premiums can be a significant cost component, and can affect the price of power and competition in the market. Moreover, the market for certain types of insurance products is thin or non-existent, and insurers' knowledge about PV systems and the PV industry is uneven. PV project developers, insurance brokers, underwriters, and other parties interviewed for this research identified specific problems with the current insurance market for PV systems in the United States and suggested government actions that could facilitate the development of this market through better testing, data collection, and communication.

Insurance premiums make up approximately 25% of a PV system's annual operating expense. Annual insurance premiums typically range from 0.25% to 0.5% of the total installed cost of a project depending on the geographic location of the installation. PV developers report that insurance costs comprise 5% to 10% of the total cost of energy from their installations, a significant sum for a capital-intensive technology with no moving parts.

Because insurance is purchased annually and future premiums are uncertain, developers who offer fixed-price contracts for the electric output of their systems must use estimated future insurance costs. Developers generally pass the risk of higher future insurance premiums on to their customers through higher escalation rates or other contract elements, thereby increasing the cost of solar electricity to entities that host PV systems on their property under power purchase agreements.

The fledgling nature of the renewable energy industry makes obtaining affordable insurance challenging. These challenges include insurers' unfamiliarity with PV technologies, a lack of historical loss data (i.e., insurance claims), and limited test data for the long-term viability of PV products under real-life conditions. The lack of information and insight about the solar PV industry contributes to perceived risk associated with the technology and installation techniques among insurance underwriters and brokers, which leads to higher premiums than would likely prevail in a more mature market. Finally, the PV industry's ongoing innovation in contractual structures and business models necessitates corresponding innovation in insurance products to match the industry's requirements.

In: Solar Photovoltaics
Editors: G. Marks and C. J. Unger

ISBN: 978-1-62257-651-7
© 2013 Nova Science Publishers, Inc.

Chapter 1

THE IMPACT OF DIFFERENT ECONOMIC PERFORMANCE METRICS ON THE PERCEIVED VALUE OF SOLAR PHOTOVOLTAICS[*]

National Renewable Energy Laboratory

LIST OF ACRONYMS

AC	alternating current
B/C ratio	benefit-to-cost ratio
CO_2	carbon dioxide
CPI	consumer price index
DC	direct current
GDP	gross domestic product
ITC	investment tax credit
IRR	internal rate of return
kW	kilowatt
kWh	kilowatt-hour
LCOE	levelized cost of energy
MACRS	modified accelerated cost recovery system
MBS	monthly bill savings

---

[*] This is an edited, reformatted and augmented version of National Renewable Energy Laboratory Technical Report, Publication No. NREL/TP-6A20-52197, dated October 2011..

MIRR	modified internal rate of return
NPV	net present value
O&M	operations and maintenance
PI	profitability index
PPA	power purchase agreement
PV	photovoltaics
REC	renewable energy certificate
SREC	solar renewable energy certificate
TNP	time-to-net-positive-cash-flow
W	watt

EXECUTIVE SUMMARY

Photovoltaic (PV) systems are installed by several types of market participants, ranging from residential customers to large-scale project developers and utilities. Each type of market participant frequently uses a different economic performance metric to characterize PV value because they are looking for different types of returns from a PV investment. We find that different economic performance metrics frequently show different price thresholds for when a PV investment becomes profitable or attractive. Additionally, several project parameters, such as financing terms, can have a significant impact on some metrics [e.g., internal rate of return (IRR), net present value (NPV), and benefit-to-cost (B/C) ratio] while having a minimal impact on other metrics (e.g., simple payback time). As such, we find that the choice of economic performance metric by different customer types can significantly shape each customer's perception of PV investment value and ultimately their adoption decision.

In this analysis, we characterize PV economic performance for three ownership types: residential customers who purchase their own PV systems, commercial customers (for-profit companies) who purchase their own PV systems, and residential and commercial customers who lease PV equipment or buy PV electricity from a third-party company. We characterize the differences in PV economics for each customer based on the different tax implications of ownership. We do not characterize PV economics for large-scale PV developers or utilities because they frequently use complex project financing structures (Harper et al. 2007) that are beyond the scope of this analysis.

We compare PV economic returns for different PV customers using the following economic performance metrics:

- Net present value (NPV)
- Profitability index (PI)
- Benefit-to-cost (B/C) ratio
- Internal rate of return (IRR)
- Modified internal rate of return (MIRR)
- Simple payback and time-to-net-positive-cash-flow (TNP) payback
- Annualized monthly bill savings (MBS)
- Levelized cost of energy (LCOE).

We characterize relative PV economics for each metric over a range of system characteristics, including PV system price and several non-price parameters including financing terms, tax rates, electricity rates and assumed rate escalations, and PV system performance. Key findings include:

- Different economic performance metrics can show unique price thresholds for when a PV investment becomes profitable or attractive.
- In some cases, the choice of an economic performance metric could have as much impact on the representation of value as decreasing (or increasing) PV prices by up to a factor of four.
- Varying non-price system characteristics, such as financing terms or assumed electricity rate increases, can impact PV economic performance as much as decreasing (or increasing) PV prices by several dollars per watt.
- At higher PV prices, commercial projects may generate higher returns than residential projects because commercial customers can depreciate the capital invested in a PV project. At lower PV prices, commercial projects may generate lower returns because the gain from capital depreciation is offset by the loss from valuing PV based on tax-deductible energy costs.
- IRR is a poor metric for characterizing the value of U.S. PV systems because the upfront nature of financed PV costs and incentives can lead to an inflated perception of value.
- MIRR and simple payback times show very little sensitivity varying several project parameters, and customers using these metrics may be less likely to be incentivized by policy measures targeting non-price system parameters.

- The MBS metric may generate attractive returns at higher PV prices than other metrics and may be effective at stimulating PV markets. However, the third-party PV companies that frequently market systems based on MBS have different tax structures and costs of capital, and it is unclear whether this will lead to higher or lower relative returns.
- The upfront nature of U.S. PV incentives (e.g., federal investment tax credit and accelerated capital depreciation for commercial customers) can lead to very different PV returns for U.S. systems relative to identical systems located in different countries that are described in the international PV literature.

The different price thresholds for when a PV investment becomes profitable or attractive and the different sensitivities to varying system parameters have significant implications for policy design. For example, if policy is introduced to improve PV financing terms, it could preferentially stimulate market segments where customers use metrics that are sensitive to financing terms (IRR, NPV, and B/C ratios) while having little or no impact on market segments where customers use metrics that are insensitive to financing terms (simple payback). There is also a strong potential for stimulating U.S. PV demand using new mechanisms in addition to traditional incentives focused on reducing PV prices or increasing revenues. These range from simply educating potential customers about the value of a PV investment as seen through different economic performance metrics, to providing access to long-term low-cost financing, to allowing third-party companies to develop simple PV products that can generate MBS.

1. INTRODUCTION

Photovoltaic (PV) systems are installed by several different types of customers. The economic returns generated by a PV investment can be very different for each market segment. This is partly caused by fundamental differences in PV prices and revenues for each market segment but can also be caused by the use of different economic performance metrics to characterize PV value. For example, a PV system may generate an internal rate of return (IRR) greater than 50%, giving a potential commercial customer the perception of a high return on investment, while an identical system could generate simple payback time that is longer than 10 years, giving a potential residential customer the perception of a low return on investment. In this analysis, we explore how the use of different economic performance metrics can shape the perception of PV value for different market participants. We calculate PV economics using several economic performance metrics for a range of PV prices to gain insight into how different metrics can exhibit unique price thresholds for when a PV investment may begin to look profitable or attractive. We also calculate the sensitivity of PV economics to non-price project parameters to gain insight into how evolving market and policy conditions could preferentially stimulate some market segments relative to others. Lastly, we highlight policy implications for the metric-dependent nature of PV economics, including unique price thresholds that may entice different PV customers to adopt PV and unique sensitivities to evolving market conditions. The report is organized as follows: Sections 2–4 describe the types of potential PV customers, the key variables that influence PV economic performance, and the economic performance metrics that are commonly used to characterize PV value; Section 5 describes the reference assumptions used to calculate PV economic performance; Sections 6–7 present results showing how PV price and non-price parameters uniquely impact each economic performance metric; Section 8 presents a method for comparing the relative sensitivity of PV economic performance across several economic metrics; Section 9 discusses policy implications; and Section 10 presents conclusions and recommendations for future research. Additional discussions of the monthly bill savings (MBS) and IRR metrics are included in Appendix A and B, respectively.

2. Types of PV Adopters and Markets

PV systems are purchased by several types of customers, and projects are frequently categorized by the type of installation:

- **Residential**—Typically roof-mounted systems that range in size from a few kilowatts up to about 10 kW. Residential PV electricity is frequently valued at retail rates and is dependent on the type of rate structure (e.g., flat rate, time-of-use rate, or tiered rate) and local net-metering policy.
- **Commercial, public sector, and non-profits**—Roof- or ground-mounted systems that range in size from a few kilowatts up to a few megawatts. Commercial PV electricity is typically valued at retail

rates and is dependent on rate structure (e.g., flat rate, time-of-use rate, or demand-based rate) and local net-metering policy.
- **Large system installers**—Typically ground-mounted arrays installed by electric-service providers or large system developers ranging in generation capacity from hundreds of kilowatts to tens of megawatts. PV electricity is frequently valued at rates set by power purchase agreements (PPAs) or by wholesale electricity market prices.

Each PV market segment has unique characteristics—including market-specific PV prices, revenues, incentives, and financing options—that affect the relative value of a PV investment. For example, residential PV prices can be twice as high on a capacity basis (installed $/W) as large-scale systems (Barbose et al. 2011; SEIA-GTM 2011). However, residential retail electricity rates ($/kWh) can be twice as high as wholesale electricity rates (EIA 2011a). In this analysis, we characterize the different tax implications for three types of PV ownership structures: (1) residential customers who own their systems, which we refer to as "residential"; (2) for-profit commercial customers, which we refer to as "commercial," who own their PV system and are not in the business of selling electricity; and (3) PV systems owned and operated by third-party companies that either lease PV equipment or sell PV electricity to residential or commercial customers. There are many other types of PV customers that we do not characterize in this analysis. These include public sector or non-profit customers that own and maintain their own systems and electric service providers (e.g., investor-owned utilities, municipal utilities, or independent electricity generators) that frequently use complex ownership structures to develop PV projects (Harper et al. 2007). While we do not characterize the full range of PV economics to each type of market participant, several system characteristics will be similar across different ownership classes. However, care must be taken in applying the general market trends evaluated here to different customer classes.

3. PARAMETERS THAT INFLUENCE PV ECONOMIC PERFORMANCE

Several project parameters influence PV economic performance. In this analysis, we evaluate how PV prices, revenues, non-price project parameters, and business models can affect PV economics and how these impacts vary depending on the use of different economic performance metrics.

PV system prices per unit of capacity ($/kW) are primarily driven by project type and size. Large PV projects can be significantly less expensive per unit of installed capacity than small PV projects, primarily because large system installers can achieve significant economies of scale. For example, average PV prices ranged from $3.85/W for utility-scale PV systems to $5.35/W for commercial systems and $6.41/W for residential systems in the first quarter of 2011 (SEIAGTM 2011). The PV prices seen by customers are also impacted by state and local incentives, which frequently target specific market segments. In this analysis, we make two simplifying cost assumptions. First, we define an "effective PV price" as the retail price minus state and local incentives[1] to generalize results. The 30% federal investment tax credit (ITC) is then applied to the remaining system cost. Second, in the base case, we assume that the reference effective PV prices and electricity rates are identical in all markets to highlight the impact of metric choice on perceived PV value. We explore the impact of varying effective PV prices and electricity rates on economic performance in Sections 6–8. All PV prices are given in units of 2010 U.S. dollars per kilowatt of direct current (DC) nameplate capacity.

Various project parameters affect PV economics in addition to system price. We consider how the following non-price parameters impact project economics and how these impacts vary depending on the use of different economic performance metrics.

- **Down payment**—The initial payment made by a potential PV customer on the debt-financed PV asset.
- **Loan term**—The duration of the PV loan, measured in years.
- **Loan rate**—The interest rate for the PV loan, given in terms of real, not nominal, rates.[2]
- **Discount rate**—The rate used to depreciate future PV revenues and costs into an equivalent present value, given in real dollars. Discount rates are frequently chosen to equal the loan rate to avoid introducing a time value of money to debt-financed capital.[3]
- **Effective tax rate**—The tax rate paid by a potential PV customer, simplified here to include federal, state, and local taxes.
- **Effective electricity rate**—The mean value of electricity generated by a PV project, measured in units of cents per kilowatt-hour. This represents a simplifying assumption intended to capture the annualized value of hourly PV generation, accounting for the daily

and seasonal variations in electricity value based on electricity prices in wholesale markets or different retail electricity rate structures (e.g., time-of-use rates based on time of day and season, demand-based rates based on peak customer power use, or tiered rates based on total energy use). Effective electricity prices could also represent rates defined in PPAs offered by a local utility.
- **Annual electricity rate increase**—The projected annual increase in electricity rates, given in units of real escalation rates.
- **Carbon price**—The projected price for emitting carbon, given in units of dollars per metric ton of carbon dioxide (CO_2) emitted.
- **Capacity factor**—The ratio of electricity generated by a PV system relative to the maximum electricity that would have been produced if the system had operated at peak capacity during an entire representative time period. PV capacity factors vary based on the local solar resource and module orientation and are given here in units of alternating current (AC) electricity generated by a given unit of nameplate DC PV capacity over one year.

There are several additional parameters that impact PV economic performance.

These include: property tax, sales tax [PV is exempt from state sales tax in several, but not all, states (DSIRE 2011)], and additional solar incentives including renewable energy certificates (RECs) and solar RECs (SRECs).

These, and other, parameters are not explicitly evaluated in this study; however, their impacts on PV economics are also metric dependent.

Historically, most PV adopters have purchased and maintained their own PV system and recouped project costs using the revenues generated by their system. However, several new business models have entered the PV market in recent years, and the different ownership structures can impact economic performance.

For example, PV systems can be owned and operated by a third-party company, which can then lease PV equipment or sell PV electricity to the building occupant (NREL 2009; Kollins et al. 2010).

PV project costs and revenues are typically taxed differently for third-party owned PV systems than customer-owned systems, which could potentially lead to higher PV returns for third-party owned systems (see Appendix A).

However, third-party companies are likely to have a higher cost of capital than customers installing their own systems. Third-party companies typically finance PV projects using several sources of capital including tax-equity investors, equity investors, and debt investors.

Most investors will require a higher rate of return than the cost of dedicated debt financing available to several residential and commercial customers.[4]

Also, the cost of capital will vary based on the third-party company, deal structure, and the PV market. For example, the cost of financing third-party residential systems may be higher than commercial systems based on increased investment risk.

4. METRICS COMMONLY USED TO REPRESENT PV ECONOMIC PERFORMANCE

Several economic performance metrics are commonly used to characterize PV value (Short et al. 1995; Duffie and Beckman 2006).

Table 1 summarizes several of these metrics. Some metric definitions, such as net present value (NPV) and IRR, are standard across different industries.

Others, such as MBS and several definitions of payback time, are specific to PV investments. Each economic performance metric characterizes PV economics in different units, including dollars, annualized percent return on investment, years, and cents per kilowatt-hour, as indicated in Table 1. Although potential PV customers might use more than one performance metric, Table 1 highlights the different PV market segments that are likely to use each metric.

NPV represents the net profit generated by an investment, calculated from the discounted sum of future costs and revenues. When the NPV of a PV system equals zero, the cost of PV-generated electricity is equal to the cost or value of electricity that could have been purchased from the grid. This is frequently referred to as reaching "grid-parity" (Denholm et al. 2009). Projects with NPVs that are greater than zero potentially represent profitable investments. NPVs generally increase with increasing revenues or decreasing costs; however, these relationships are not always intuitive because the relative timing of project costs and revenues is important since they are discounted. NPVs cannot, by themselves, be used to rank the relative returns of investments with different costs. To compare between different investments, NPVs can be scaled by the investment cost, which results in the profitability index (PI) metric, or variations of NPV can be calculated like the benefit-to-cost (B/C) ratio. The NPV metric is likely to be used to evaluate commercial and large-scale PV systems and possibly some residential PV systems.

Table 1. Metrics Used to Characterize PV Economic Performance

Metric	Equation	Units	Likely User
Net Present Value (NPV)	$NPV = \sum_{t=0}^{N} \dfrac{Revenue_t - Cost_t}{(1+d)^t}$	$	Some residential Commercial Large-scale
Profitability Index (PI)	$PI = \dfrac{\sum_{t=0}^{N} \dfrac{Revenue_t - Cost_t}{(1+d)^t}}{Investment\ Cost}$	%	Some residential Commercial Large-scale
Benefit-to-Cost (B/C) Ratio	$B/C\ Ratio = \dfrac{\sum_{t=0}^{N} \dfrac{Revenue_t}{(1+d)^t}}{\sum_{t=0}^{N} \dfrac{Cost_t}{(1+d)^t}}$	%	Commercial Large-scale Public sector
Internal Rate of Return (IRR)[b]	$IRR: NPV = \sum_{t=0}^{N} \dfrac{Revenue_t - Cost_t}{(1+IRR)^t} = 0$	%	Some residential Commercial Large-scale
Modified Internal Rate of Return (MIRR)	$MIRR = \left(\dfrac{\sum_{t=1}^{N} PositiveCashFlow_t * (1+r)^{N-t}}{\sum_{t=1}^{N} \dfrac{NegativeCashFlow_t}{(1+d)^t}} \right)^{\frac{1}{N}} - 1$ $PositiveCashFlow_t = \begin{cases} Revenue_t - Cost_t & if\ Revenue_t > Cost_t \\ 0 & if\ Revenue_t \leq Cost_t \end{cases}$ $NegativeCashFlow_t = \begin{cases} 0 & if\ Revenue_t \geq Cost_t \\ Revenue_t - Cost_t & if\ Revenue_t < Cost_t \end{cases}$	%	Some residential Commercial Large-scale

The Impact of Different Economic Performance Metrics …

Metric	Equation	Units	Likely User
Payback Time[c]	$$\text{Simple Payback} = \frac{PV\ Price - Federal\ ITC}{Annual\ PV\ Revenue - O\&M}$$ $$\text{TNP Payback}: \sum_{t=0}^{TNP\ Payback} \frac{Revenue_t - Cost_t}{(1+d)^t} > 0\ \&$$ $$\sum_{t=TNP\ Payback}^{N} \frac{Revenue_t - Cost_t}{(1+d)^t} > 0$$ Several others (e.g., Duffie and Beckman 2006)	years	Residential Some commercial
Annualized Monthly Bill Savings (MBS)[d]	$$MBS = \frac{1}{LeaseTerm * 12} \sum_{t=1}^{LeaseTerm} \frac{PV\ Generation_t * (Electricity\ Rate_t - LCOE)}{(1+d)^t}$$	$/month	Residential Some commercial
Levelized Cost of Energy (LCOE)[e]	$$LCOE_{Residential} = \frac{\sum_{t=0}^{N} \frac{Cost_t}{(1+d)^t}}{\sum_{t=0}^{N} \frac{Electrical\ Energy_t}{(1+d)^t}}$$ $$LCOE_{Commercial} = \frac{\sum_{t=0}^{N} \frac{Cost_t}{(1+d)^t}}{\sum_{t=0}^{N} \frac{Electrical\ Energy_t}{(1+d)^t}} * \frac{1}{(1-Commercial\ Tax\ Rate)}$$	cents/kWh	Primarily large-scale

Note: N represents the number of years for the economic analysis; t represents the year variable in each summation; d represents the discount rate, which we have also used interchangeably with the loan rate; $Revenue_t$ represents the revenue generated by the PV system in year t; $Cost_t$ represents the cost of the system in year t; and r in the MIRR formulation represents the reinvestment rate, which is a company's opportunity cost of capital. All other variables have descriptive labels.

[a] In this analysis, we use the term "commercial" to represent for-profit commercial entities that pay taxes.

[b] See Appendix B for further discussion.

[c] TNP payback is defined as the time required to satisfy two conditions: (1) the discounted PV revenues exceed the discounted system costs accrued to that date and (2) the discounted revenues remain higher than discounted costs for the duration of the investment

[d] See Appendix A for further discussion.

[e] Represents LCOE in real, not nominal, dollars. Commercial LCOEs are adjusted to represent the before-tax cost of electricity that can be compared to retail or wholesale electricity rates.

The PI represents the project NPV divided by the initial investment cost. PIs represent the discounted percent return on an investment, and PIs greater than zero represent profitable investments. Since PIs are normalized by the investment price, they can be used to rank the relative returns from several investments with different costs.[5] The B/C ratio represents the discounted system revenues divided by the discounted system costs. A B/C ratio greater than one represents a profitable investment. The main difference between the PI and B/C ratio is that all costs in the B/C ratio are discounted, whereas PI is calculated by normalizing the difference between discounted revenues minus costs (NPV) by the undiscounted initial investment cost. These differences are generally small, and PI frequently shows similar returns as the B/C ratio.

Both metrics are likely to be used by potential commercial and large-scale PV market segments and may additionally be used by some residential customers. The B/C ratio is frequently used in the public sector.

The IRR represents the discount rate at which the project NPV equals zero and is frequently interpreted as the annualized return on investment. The upfront nature of PV costs and tax incentives can lead to several challenges in calculating and interpreting PV IRRs, which are discussed in detail in Section 7 and Appendix B. Modified IRRs (MIRRs) are similar to IRRs, but positive net revenues are explicitly reinvested at the company's, or an individual's, opportunity cost of capital[6] rather than implicitly reinvested at a rate equal to the system IRR. This tends to shift low IRRs up to the reinvestment rate and high IRRs down to the reinvestment rate (McKinsey & Co. 2004). The IRR and MIRR metrics are likely to be used to evaluate commercial and large-scale PV investments.

Investment payback times have several definitions (Duffie and Beckman 2006). We include two in this analysis:

- **Simple payback time**—The time required for undiscounted PV net revenues to equal the initial investment cost (Perez et al. 2004; Paidipati et al. 2008; Black 2009).
- **Time-to-net-positive-cash-flow (TNP) payback time**—The time required for: (1) the discounted PV revenues to exceed the discounted system costs accrued to that date and (2) the discounted revenues to remain higher than discounted costs for the duration of the investment (Nofuentes et al. 2002; Sidiras and Koukios 2005; Audenaert et al. 2010).

Simple and TNP payback times are the most frequently used payback metrics for PV investments, but there are several other payback definitions. Although we do not evaluate the relative economics of each payback definition, the difference between simple payback times and TNP payback times illustrate the wide range in payback times that are generated by different payback definitions. These differences are primarily driven by the fact that simple payback times are not sensitive to financing parameters or the relative timing of system costs and revenues, whereas other payback metrics can be very sensitive to these and other parameters. The simple and TNP payback metrics are likely to be used to evaluate residential and some commercial PV investments.

Annualized MBS are used to characterize the potential average decrease in a PV customer's electricity bill resulting from a PV investment. We estimate this based on the difference between PV levelized costs of energy (LCOEs) and effective electricity rates, multiplied by the discounted electricity generated by a PV system in a given year. Based on differences in PV electricity generation profiles and retail electricity rates in different months, the MBS in any given month will likely be higher or lower than the annualized MBS.

The MBS metric is frequently used by third-party PV companies to characterize PV value to potential customers (NREL 2009; SolarCity 2011; SunRun 2011). Third-party PV companies can repackage PV costs and revenues into a simple product that shows monthly bill savings, and their customers are likely to characterize PV value in these terms. This potentially reduces the complexity of valuing a PV investment by framing PV returns in an intuitive measure that may be more likely to entice customer adoption (Wilson and Dowlatabadi 2007). While a mean annualized MBS can be calculated for customer-owned systems, the annualized MBS will not reflect the actual PV costs and revenues generated by a PV investment, and it is less likely that customer-owned PV adopters will use MBS to represent investment value.

The MBS earned by a PV project can vary for different ownership structures. For example, a third-party owned residential PV system represents a depreciable asset, which could potentially allow a third-party owned PV system to produce higher bill savings than a residential customer. The difference for commercial systems is not as pronounced because commercial customers already depreciate PV assets, but third-party owned MBSs could potentially be slightly higher, as described in Appendix A. However, the higher cost of capital for third-party PV companies will reduce potential PV returns, and it is unclear whether the combination of different tax structures and costs of capital will lead to higher or lower returns. We explore the impact ownership-dependent tax structures, but not costs of capital, in this analysis.

The LCOE represents the discounted price that PV electricity must be sold at to recoup discounted project costs over the life of the system. PV LCOEs are calculated in units of real dollars in this analysis, whereas PV project developers and utilities often use nominal LCOEs. Unlike all other economic performance metrics in this analysis, PV LCOEs are relative metrics that must be compared to the value of the electricity generated, which can range from the electricity price seen by the customer or LCOEs from different technologies. These valuations frequently do not capture the general correspondence of PV generation with times of peak electricity demand and electricity prices (Borenstein 2008), and the use of different comparison values can lead to a wide range in the perceived economics of identical PV systems. Other metrics may be less prone to multiple interpretations of value.

PV users frequently use different economic performance metrics because they prioritize PV investment risk and returns differently. For example, home owners might be interested in PV systems with short payback times because they are uncertain about how long they will live in their current home and how a PV investment will affect their home's value. Research has suggested that residential customers, and some commercial customers, are more likely to use payback times to characterize the value of a PV investment or other energy-saving investments (Kastovich et al. 1982; Perez et al. 2004; Sidiras and Koukios 2005; Black 2009). Residential and commercial customers may also think of PV value in terms of how much their monthly electricity bills will decrease if they invest in PV, and third-party owned PV companies frequently market PV products using bill savings metrics (NREL 2009; SolarCity 2011; SunRun 2011). Potential commercial PV customers may think of PV as a longer-term investment than residential customers and may be more likely to characterize PV value as an annualized return on investment (Chabot 1998; Talavera et al. 2007; Talavera et al. 2010). Commercial customers may use B/C ratios, PIs, IRRs, or MIRRs to compare potential PV returns relative to other investment opportunities. Utilities and large-scale developers frequently characterize PV costs in terms of LCOE (CEC 2007; SunPower Corp. 2008), which can be compared to wholesale electricity prices, local PPA offerings, or the LCOEs of different generation technologies. Large developers may also use additional metrics such as the B/C ratio, NPV, IRR, MIRR, or others to rank PV investment performance relative to other investment opportunities.

Several recent European studies have recommended using IRRs to characterize PV value (Nofuentes et al. 2002; Talavera et al. 2007; Talavera et al. 2010; Audenaert et al. 2010). Wind investors frequently use IRR-based hurdle rates[7] to characterize project economics (Harper et al. 2007) and may apply similar investment criteria to evaluate solar projects. Others have recommended the use of metrics based on NPV, such as PI (Chabot 1998; Nofuentes et al. 2002; Audenaert et al. 2010).

5. REFERENCE PV SYSTEM ASSUMPTIONS

To calculate the sensitivity of PV economic performance to a range of system parameters, we first define reference PV price, performance, and financing assumptions and apply these to all PV market participants (Table 2). We assume identical system characteristics for each market participant, which does not capture the differences seen for each customer type.[8] However, this assumption is made to highlight how the use of different economic performance metrics can shape the perceived value of a PV investment.

The reference assumptions in Table 2 are used to evaluate the relative economics for residential and commercial systems for each economic performance metric over a range of price, performance, and financing parameters. Some states have a combination of PV incentives and retail electricity rates where PV economic performance may meet or exceed the reference conditions (e.g., Florida, Hawaii, New York, New Jersey, and parts of California) (DSIRE 2011; EIA 2011b).

PV systems installed in other states may generate lower economic returns than the reference conditions. One challenge in characterizing the value of PV electricity is that several rate structures vary by time or season (time-of-use rates), peak electricity use (demand-based rates), or total electricity use (tiered rates). Regardless, the reference values in Table 2 are not meant to characterize representative U.S. PV economic performance; they are meant only as a starting point for the sensitivity analysis, which is used to compare the relative value of PV, as shown by different economic performance metrics. Figure 1 shows annual after-tax PV cash flows for customer-owned residential and commercial PV systems that are calculated using the reference parameters in Table 2. PV costs are primarily composed of an initial down payment, followed by annual loan payments and operation and maintenance (O&M) costs.

Table 2. Reference PV System Parameters

	Reference Characteristics
Effective PV Price[a]	$4,000/kW
Capacity Factor[b]	17%
Annualized Electricity Rate[c]	$0.15/kWh
Annual Electricity Rate Increase (real dollars)	None
PV Performance Degradation	0.5%/year
Down Payment	20%
Loan Rate (real)	5%
Loan Term	20 years
Duration of Economic Analysis	30 years
Capital Reinvestment Rate[d] **(real)**	8%
Discount Rate[e] **(real)**	5%
Incentives	30% federal ITC; capital depreciation for commercial customers[f]
Net Metering	Full
Carbon Policy	None[g]
Annualized Operations and Maintenance Payment[h]	$35/year for years 1-10 $25/yr for years 11-20 $20/yr for years 21-30
Analysis Term	30 years
Tax Rates	State and federal tax rates are combined into an aggregate tax rate of 35% for residential customers and 40% for commercial customers
Tax Implications	After-tax energy costs for residential; before-tax energy costs for commercial

[a] Effective PV price (in 2010 U.S. dollars) represents the system price after taking state and local PV incentives but not the 30% federal ITC.

[b] A 17% PV capacity factor roughly represents PV output from a fixed-tilt (tilt = latitude) residential PV system in Kansas City, Missouri (SAM 2011). Similar PV systems are likely to perform better (21.5% capacity factor in Phoenix, Arizona) or worse (15.5% capacity factor in Chicago, Illinois) (SAM 2011).

[c] The reference annualized rate is higher than the average U.S. retail electricity rate from June 2010 for residential ($0.12/kWh) and commercial ($0.11/kWh) customers (EIA 2011b). However, the annualized rate is at or below the June mean electricity rates for states in New England, the Middle Atlantic, and California (EIA 2011b).

[d] The capital reinvestment rate is used to calculate the MIRR and represents the rate of return a company could receive on investments of similar risk outside the project. This is frequently referred to as the opportunity cost of capital (Stermole and Stermole 2009).

[e] The discount rate is assumed to be the same as the loan rate in the reference scenario to avoid introducing a time value of money for debt financed capital.

[f] Commercial depreciation follows a five year Modified Accelerated Cost Recovery System schedule (DSIRE 2011).

[g] We do not include carbon policy in the reference scenario; however, we do include carbon prices in the sensitivity analysis. We assume a 0.58 kg CO_2/kWh carbon intensity, based on mean emissions rates from the U.S. electricity sector (EIA 2011a), to represent the increase in electricity rates for a range of carbon policies.

[h] Operation and maintenance costs are assumed to decrease over time to reflect technical improvements, primarily longer inverter lifetimes, and lower costs.

These costs are partially offset by system tax benefits, including the federal ITC, Modified Accelerated Cost Recover System (MACRS) capital depreciation for commercial customers, and tax-deductible payments on loan interest.[9]

Annual PV revenues are primarily based on the PV output multiplied by the annualized value of PV electricity. PV revenues are also affected by state and local net-metering policy (NNEC 2010).[10] PV generation decreases over time based on an assumed 0.5%/year system degradation, with a corresponding decrease in annual PV revenue.

Figure 1 shows that the largest annual PV costs (down payment) and tax incentives (federal ITC and MACRS) occur in the first few years of system ownership. After this, the reference PV costs and revenues are nearly identical, leading to small net revenues or net costs each year. The upfront nature of PV costs and incentives has a significant impact on some economic performance metrics (e.g., IRR and TNP payback) but not others (e.g., simple payback).

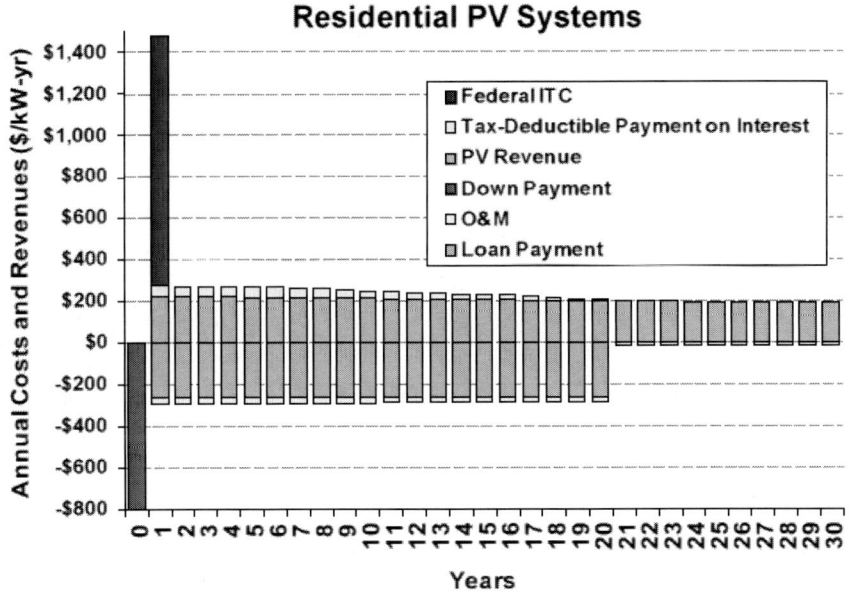

a

Figure 1. (Continued)

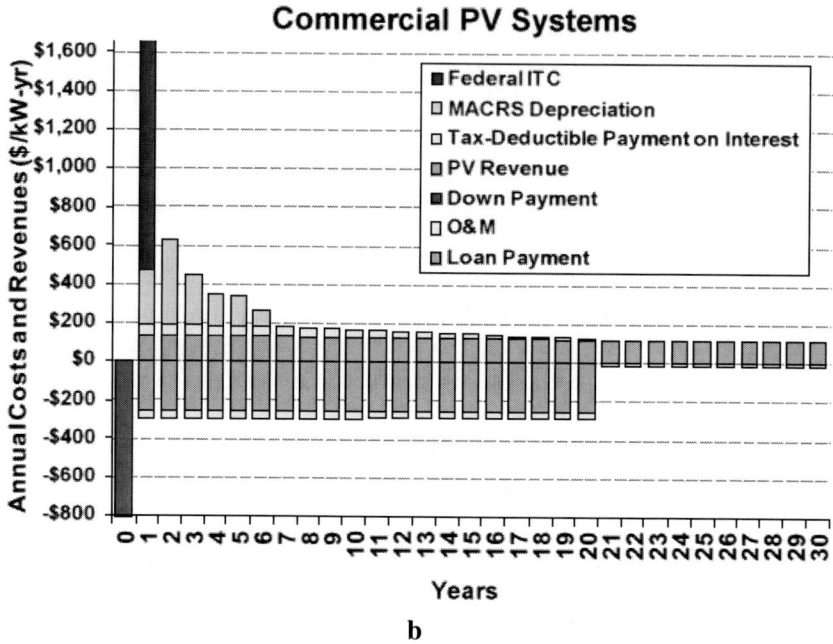

Figure 1. Reference residential (a) and commercial (b) undiscounted annual PV system costs, revenues, and tax benefits given in real dollars.

In addition to the customer-owned PV systems shown in Figure 1, we also characterize the MBS that could be offered by third-party owned PV companies (see Section 4.5 and Appendix A) using the project assumptions listed in Table 2. Third-party owned PV systems are a rapidly growing market segment (Drury et al. 2011; SEIA-GTM 2011) because of their potential to reduce several adoption barriers like upfront adoption costs and technology risk and complexity. We focus the analysis only on the potential MBS that could be offered to a building occupant, and we do not characterize PV returns to third-party companies or various tax, equity, and debt investors. We also assume the same PV prices, and other system parameters listed in Table 2, as those used for other ownership structures.

The following sections evaluate the sensitivity of PV economic performance to varying system price (Section 6) and non-price (Section 7) system characteristics. The sensitivity of PV economic performance to varying prices provides insight into how different economic metrics can exhibit unique price thresholds for when a PV investment may begin to look attractive. The sensitivity of PV economic performance to each non-price parameter (PV generation, financing, and electricity market assumptions) provides insight into how future market and policy projections could differentially impact PV market participants.

6. SENSITIVITY ANALYSIS—EFFECTS OF PV PRICE

In this section, we evaluate the relative sensitivity of PV economic performance to a range of system prices and evaluate the different thresholds for when PV becomes a profitable investment for different performance metrics. Figure 2 shows PV economic performance for a range of effective PV prices from $1,000–$7,000/kW,[11] calculated for each economic performance metric. Effective PV prices represent the total installed system price after taking state and local incentives but before taking the 30% federal ITC.[12] This range in effective PV prices covers the range of current PV prices seen by U.S. customers, which are subject to widely varying state and local PV incentives (DSIRE 2011). All non-price assumptions are based on the reference parameters in Table 2, and the economic returns represent after-tax valuation for both residential and commercial systems.[13] As such, the returns from a PV investment must be compared to the after-tax returns from other investment opportunities. Because energy costs are tax deductible for commercial customers, commercial LCOEs are adjusted to represent the before-tax cost of electricity to compare with retail or wholesale electricity rates. Both the NPV and MBS represent the value generated by 1 kW of PV capacity; the actual net savings or costs will be larger based on system size. Lastly, we add one to the PI metric (1 + PI) so that PI results can be compared to B/C ratios. This scaling does not affect PI sensitivities—it just shifts the results so that returns are positive if (1 + PI) is greater than one, and returns are negative if the results are less than one.

Figure 2 shows the different sensitivities of PV economic performance to price. The behavior of different metrics can be roughly categorized as those showing: (1) a nearly linear response to changing PV prices, including NPV, MBS, MIRR, simple payback times, and LCOE; (2) a nonlinear, smoothly varying response to changing prices, including B/C ratio and PI; and (3) strong threshold behavior, including IRR and TNP payback times. Potential customers can typically expect a higher sensitivity to changing PV prices when using metrics that show non-linear or threshold behavior. For example, IRR shows a 4.5% annualized return on a $4,800/kW residential PV system and a 30.3% return on a $4,400/kW system.

The decrease in effective PV price from $4,800/kW to $4,400/kW could potentially make a PV investment look attractive to a residential customer if they use IRR to characterize PV value. However, this same decrease in residential system prices leads to a $0.01/kWh decrease in residential PV LCOE (from $0.15/kWh to $0.14/kWh), a two-year reduction in simple payback time (from 23.9 to 21.9 years), a $0.70/kW-month increase in MBS for customer-owned systems (from −$0.15/kWmonth to $0.55/kW-month), and about a 5% increase for the B/C ratio (from 0.99 to 1.04) and PI metric (from −0.01 to 0.04).

The improvement in IRR returns for a reduction in system price from $4,800/kW to $4,400/kW is much larger than the associated improvements for other metrics and customers using IRR may be more likely to adopt PV given this price reduction. Commercial PV systems show similar non-linear and threshold behavior to changing prices.

Both commercial and residential IRRs show very high returns relative to other metrics and strong threshold behavior. This is primarily caused by the upfront nature of system costs (down payment) and incentives (federal ITC and MACRS for commercial customers). The timing of system costs and revenues leads to several challenges for calculating and interpreting PV IRRs, and we generally find that IRR is a poor metric for characterizing the value of U.S. PV systems, as discussed in Appendix B.

One consequence of the metric-dependent nature of PV returns is that each metric can show a different price threshold for when a PV investment becomes profitable or attractive enough to entice adoption. For example, residential NPV, PI, and MBS (customer owned) all pass from negative to positive returns when PV prices reach $4,700/kW. At this same price, residential PV LCOEs become less expensive than effective electricity rates, and the B/C ratio passes through unity, representing a profitable investment. However, residential IRRs become positive at $5,400/kW, and MIRRs are positive for all system prices but do not exceed the assumed 8% (real) reinvestment rate[14] until PV prices are below $4,200/kW. Third-party owned MBS are also positive for all system prices evaluated in Figure 2 (Appendix A). Simple payback times are about 23 years for $4,700/kW residential PV systems and are not reduced to less than 10 years until effective PV prices reach $2,000/kW. Residential TNP payback times transition from more than 20 years to less than 2 years at $3,800/kW.

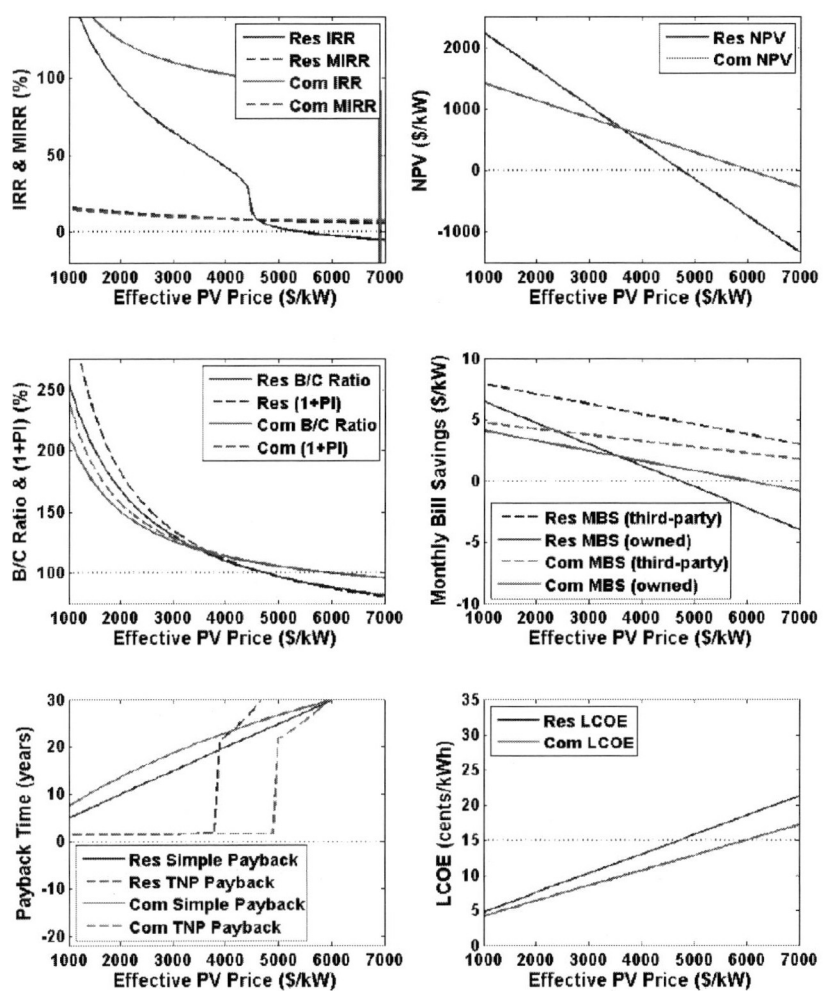

Note: Here and elsewhere, PI is shifted by adding one (1 + PI) to better compare PI performance to B/C ratios. MBSs are shown for both customer-owned and third-party owned systems to characterize the different tax implications of ownership. The dotted black line shows the transition from an unprofitable to a profitable investment, with the exception of payback time, where it shows an instantaneous payback that defines the lower limit. All returns are after-tax and given in units of real dollars.

Figure 2. PV economic performance, characterized using several metrics, for a range of effective PV prices for residential ("Res") and commercial ("Com") systems.

We find similar relationships for commercial PV systems, where a $6,000/kW system price will produce positive NPVs, PIs, and MBSs (customer owned), while different metrics can show higher or lower price thresholds for when an investment becomes profitable. Commercial systems typically show higher returns than residential systems at higher PV prices and lower relative returns at lower prices, as shown by the lower slope between commercial PV returns and price relative to residential systems. This is because, at higher PV costs, MACRS depreciation typically has a greater impact on net returns than the decrease in commercial PV revenues (commercial energy costs are tax deductible). The converse is true at lower PV prices, where the reduction in commercial PV revenues has a greater impact on project economics than MACRS depreciation. Not all metrics follow this general trend. For example, payback times are typically lower for residential systems than commercial systems.

The difference in PV price thresholds between metrics could significantly shape the perceived value of a PV investment for different customers. For example, a commercial customer may be interested in investing in a $6,000/kW PV system if they use positive system NPVs, PI, and B/C ratios as their investment criteria. However, a similar commercial customer may base their investment decision on achieving a simple payback time of 10 years or less and wait for effective PV prices to reach $1,300/kW before investing. In this case, the choice of an economic performance metric could have as much impact on the investment decision as changing PV prices by a factor of four. In addition, even if residential and commercial customers used the same investment criteria (e.g., a positive NPV or PI), the commercial customer may be enticed to adopt a $6,000/kW system while a residential customer may wait for effective PV prices to reach $4,700/kW.

Another challenge is to understand not only what price thresholds represent a *profitable* investment (net revenues exceed costs) but what prices represent an *attractive* investment (returns are high enough to entice customer adoption). For example, a PV system that generates a PI greater than zero represents a profitable investment, but a potential customer may look for additional returns such as a PI greater than 0.2 before investing (Chabot 1998). However, the system that produced a PI greater than zero would generate a positive MBS, which may be sufficient to entice customer adoption. In this case, the system characteristics are identical and the returns are similar, but a customer may perceive the value of PV to be higher if returns are described in terms of bill savings rather than PI, which could lead to higher effective price thresholds for adoption. This is not specific to PV; general customer behavior has been shown to be influenced by the way information is presented (e.g., Wilson and Dowlatabadi 2007), and customers typically value near-term savings or costs much more than savings or costs that occur in the distant future (see Section 9).

Lastly, Figure 2 shows that third-party owned PV systems could potentially generate higher MBSs than identical customer-owned systems based on the different tax structures for third-party ownership (see Appendix A). The higher returns are based on the assumption that third-party PV companies can depreciate PV assets based on MACRS and that they are likely to have a higher tax rate, which typically increases PV value (see Section 7). Because of this, a third-party owned residential and commercial system could potentially generate positive bill savings for a $7,000/kW system, while an identical customer-owned residential or commercial system would have to reach $4,700/kW or $6,000/kW, respectively, to generate positive bill savings. The MBS for third-party owned PV systems show profitable returns at higher costs than any other metric. However, we assume the same cost of capital for third-party owned systems and customer-owned systems, which likely overestimates the potential MBS that could be offered by third-party PV companies.

7. SENSITIVITY ANALYSIS—EFFECTS OF NON-PRICE PARAMETERS

In this section, we evaluate the relative sensitivity of PV economic performance to a range of non-price system parameters for several performance metrics: PI[15] and B/C ratio, payback times, MBS, IRR and MIRR, and LCOE. We calculate and compare the relative sensitivities for: (1) several non-price parameters within one metric and (2) one non-price parameter between several metrics. The relative sensitivity of one metric to several non-price parameters gives insight on how a market participant, using a specific metric, may react to varying system parameters, either through a natural evolution of market conditions or through directed policy. The relative sensitivity of varying one non-price parameter across several metrics gives us insight into how a market or policy change could differentially impact several different types of market participants.

Figure 3 shows the sensitivity of the B/C ratio and PI to a range of non-price system characteristics for a PV system at the reference $4,000/kW[16] price. We shift PI by one (1 + PI) in Figure 3 to better compare PI and B/C ratio sensitivities. In general, the B/C ratio and PI show similar returns and sensitivities, are smoothly varying, and do not exhibit threshold behavior.

The B/C ratio and PI sensitivities illustrate several general trends. PV returns typically increase if: (1) the system costs are spread out over more years (lower down payment fraction or longer loan term), (2) the system costs are reduced (lower loan rates[17] or higher tax rates), or (3) the system revenues are increased (increased electricity rates, positive rate escalations, carbon emission pricing,[18] or increased capacity factors). We find that B/C ratios and PIs show similar returns, and the use of either metric is roughly interchangeable with the exception of varying discount rates.[19]

Several metrics show higher commercial returns and lower residential returns for increasing discount rates. For commercial customers, this trend is based on the increased impact of near-term incentives (30% federal ITC and MACRS) and the decreased impact of later-term cash flows where loan payments and O&M costs often exceed PV revenues, as shown in Figure 1. The converse is true for residential systems, where the increased influence of near-term incentives (30% federal ITC) has less impact than the reduced influence of later cash flows where PV revenues often exceed loan payments and O&M costs.

The B/C ratio, PI, and several other metrics frequently show higher returns for increasing tax rates since higher tax rates effectively lower the price of a financed PV system. This is because payments on loan interest are typically tax deductible, which effectively reduces the cost of a PV system. Commercial customers can also depreciate a PV asset following MACRS, which increases in direct proportion to the company's tax rate. However, commercial PV revenues also decrease in direct proportion to their tax rate, and we find the benefits from MACRS depreciation are roughly canceled by the decrease in revenues for a $4,000/kW PV system. While we do not calculate PV economics for public sector and non-profit entities, the economics of these systems are shown for reference parameters by a tax rate of zero. PV returns are significantly lower for these systems, and states frequently compensate for this by developing larger incentives for this market segment (DSIRE 2011). Another option for the public sector or non-profits could be to adopt third-party owned PV systems (Bolinger 2009), where the third-party company could benefit from non-zero tax rates and potentially pass these benefits on to the end user.

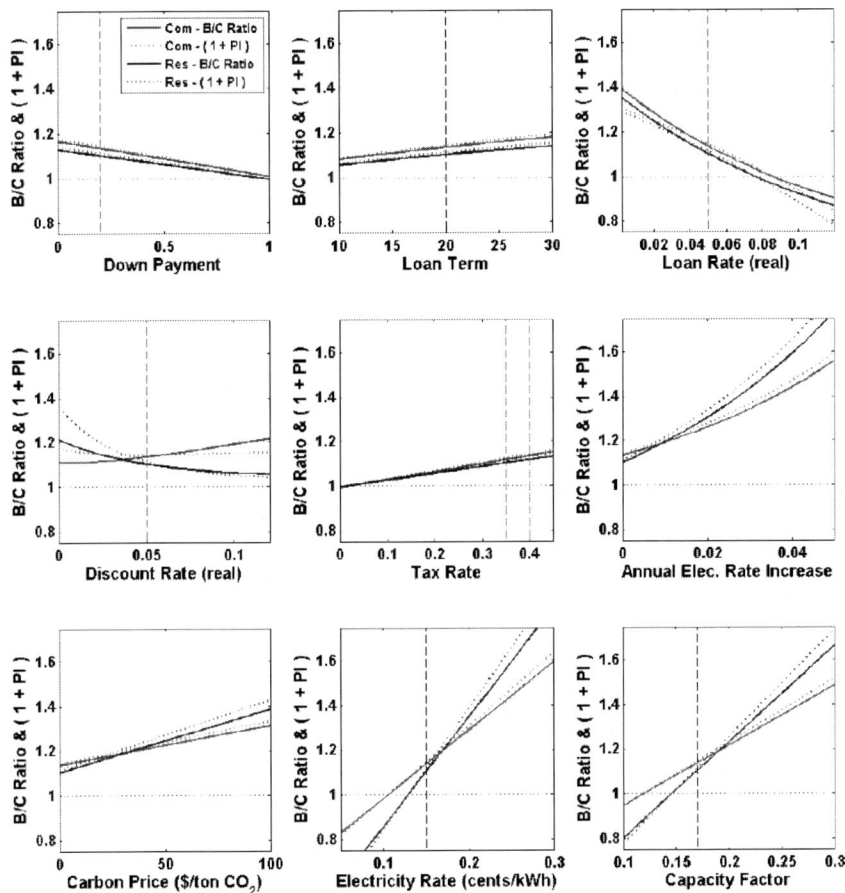

Note: Vertical dashed lines show the reference PV assumptions, and horizontal dotted grey lines show the demarcation between a profitable and unprofitable investment.

Figure 3. B/C ratios and the shifted PI (1 + PI) are shown for $4,000/kW residential and commercial PV systems over a range of financing, performance, and market parameters.

PV returns also predictably increase with increasing system revenues. PV returns can be particularly sensitive to the assumed increase in electricity rates over time, and PV retailers often assume a non-zero increase in real electricity rates over time when characterizing PV economic returns (e.g., SolarCity 2011; Sun Light & Power 2011). For example, a 3% annual electricity rate increase (real dollars) has as much of an impact on residential system economics as decreasing the loan rate to about 0% or introducing a $100 per metric ton of CO_2 price. Rate increases are also frequently written into PPA contracts offered by third-party PV owners, with a similar increase in the returns third-party PV providers generate from their PV assets. Potential PV customers should be careful to understand how this and other market assumptions affect PV economics, and this represents a strong opportunity for the public sector to help inform potential PV adopters.

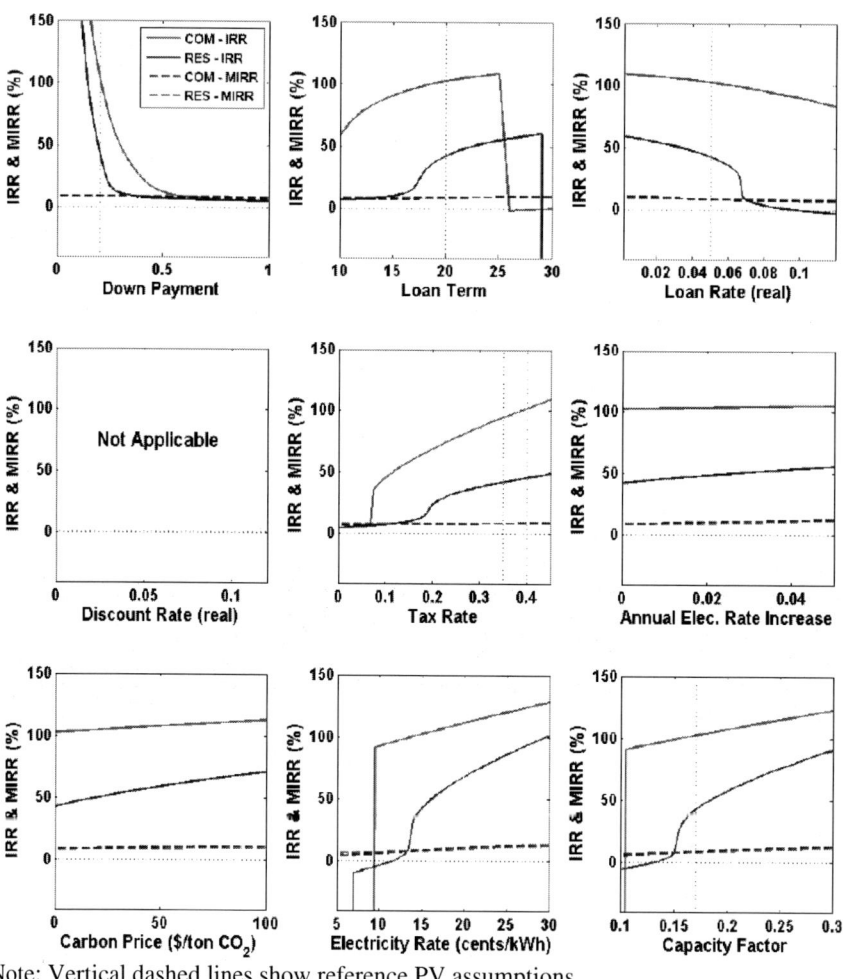

Note: Vertical dashed lines show reference PV assumptions.

Figure 4. IRR and MIRR for $4,000/kW residential and commercial PV systems over a range of financing, performance, and market parameters.

Figure 4 shows IRR and MIRR for residential and commercial PV customers for a range of non-price system characteristics. IRR is very high for the reference system parameters (42% for residential and 102% for commercial) and exhibits strong threshold behavior. These trends are driven by the upfront nature of PV costs (loan down payment) and revenues (federal ITC and MACRS for commercial customers). IRRs are also sensitive to cash flows that change sign, and net PV cash flows frequently oscillate from negative to positive multiple times, as discussed in Appendix B.

PV IRRs are very sensitive to variables that affect the timing of system costs (down payment fraction and loan term) and upfront incentives (tax rate that scales MACRS depreciation and tax-deductible payments on loan interest). Commercial IRRs are typically much higher than residential IRRs because the upfront nature of MACRS depreciation has more of an impact on system IRRs than the reduced PV revenues (energy costs are tax deductible for the commercial customers considered here). The IRR is less sensitive to parameters that affect mean system costs (loan rate) or revenues (electricity rate escalations and carbon prices) unless the variations are sufficient to move IRR beyond a threshold (electricity rates and capacity factors).

When PV cash flows are sufficient to achieve system IRRs of about 50% or higher, IRRs become relatively insensitive to further improvements in system characteristics. This is because high IRR solutions represent very high discount rates, which magnify the importance of PV costs and revenues in the first few years of ownership and reduce the impacts of costs and revenues in later years. For example, pricing carbon emissions would lead to higher PV returns over the life of the investment, but for a commercial system with an IRR (and discount rate) greater than 100%, the benefits are so highly discounted that IRR becomes virtually insensitive to carbon price.

The IRR is very sensitive to the loan down payment fraction. This is primarily because it impacts the balance of upfront costs (loan down payment) and incentives (federal ITC and MACRS depreciation). The sensitivity to down payment fraction is important for residential customers who may pay for the system out of pocket (corresponding to a 100% down payment), finance the system through a home-equity loan (zero or small down payment), or roll the system cost into the home mortgage for new homes, with a corresponding wide range in potential down payment fractions. The sensitivity to down payment fraction is also important for commercial customers. Small commercial companies may get dedicated financing for a PV investment, while larger commercial customers may use existing capital and debt reserves to develop the system. If the latter method is used, the down payment fraction is roughly based on a company's debt-to-equity ratio, which can vary significantly both within and across industries. Since PV IRRs are so dependent on this uncertain parameter, it is challenging to specify a reference IRR for each type of market participant. Rather, PV economic analysis should be calculated on a project-specific basis.

We generally find higher IRRs for U.S. PV systems than the IRRs calculated for similar European PV systems (Audenaert et al. 2010; Talavera et al. 2010). This difference is driven by the upfront incentives available to U.S. systems (federal ITC and MACRS depreciation for commercial systems), as compared to the production-based incentives frequently available in European PV markets. We also find that the IRRs calculated for U.S. PV systems are much more sensitive to the customer's tax rate than the IRRs calculated for European systems (Talavera et al. 2010) because tax rates directly scale both MACRS depreciation and the tax-deductible payments on loan interest for U.S. systems. Generally, IRRs for U.S. systems also exhibit stronger threshold behavior because of the upfront nature of PV costs and incentives. Unlike recent European studies, we find that IRR is a poor measure of PV value because of the differences between U.S. versus European incentives. We evaluate the challenge of calculating and interpreting PV IRRs for U.S. systems in further detail in Appendix B.

The MIRR has been proposed as a better metric for characterizing investment returns than the IRR (McKinsey & Co. 2004). However, the upfront nature of PV costs and incentives makes the MIRR highly sensitive to the assumed reinvestment rate (8% in the reference case) and relatively insensitive to other system characteristics. For example, the reference commercial MIRR is 8.7%. This MIRR increases to 9.9% if the annualized electricity rate is increased from $0.15/kWh to $0.20/kWh and increases to 12.6% if the electricity rate is increased to $0.30/kWh. MIRR also shows a similarly small increase for decreasing PV prices (Figure 2). These and other changes in PV price and performance characteristics have a far greater impact on the other economic performance metrics. The upfront nature of PV costs and incentives makes the MIRR unresponsive to varying input parameters and strongly dependent on the assumed reinvestment rate, and we generally find that the MIRR is a poor measure of PV value for U.S. systems.

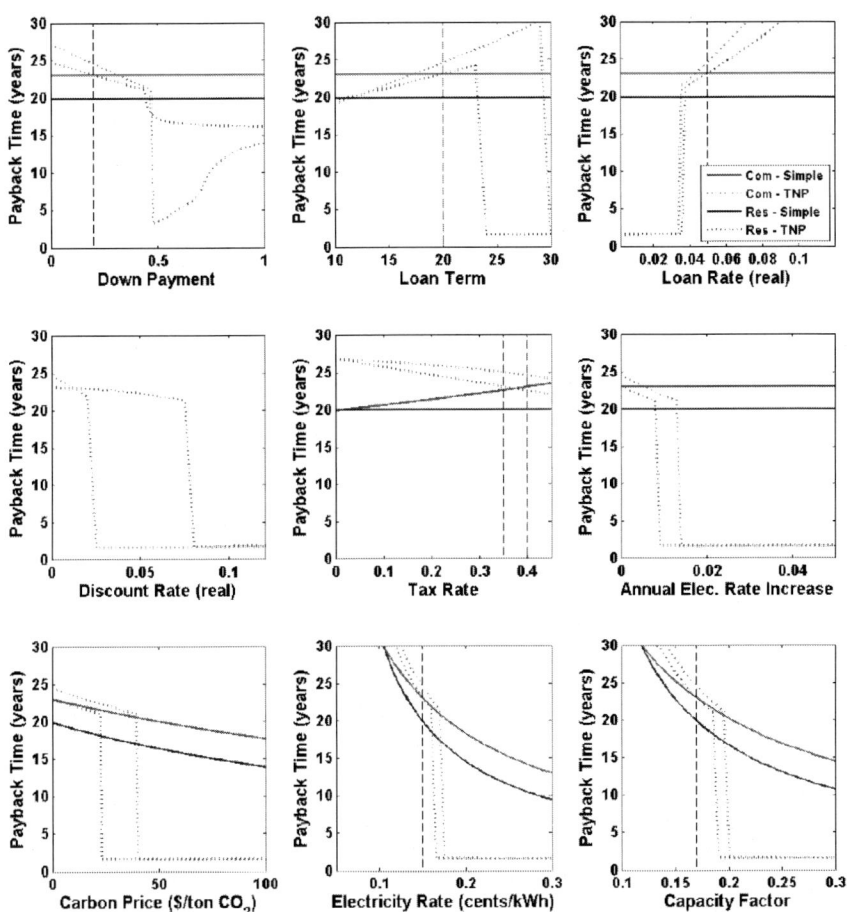

Note: Vertical dashed lines show reference PV assumptions.

Figure 5. Simple and TNP payback times for $4,000/kW residential and commercial PV systems over a range of financing, performance, and market parameters.

Potential residential and smaller commercial customers may use payback times to characterize the value of PV or other energy efficiency investments (Kastovich et al. 1982; Perez et al. 2004; Sidiras and Koukios 2005). One challenge is that there are several definitions of PV payback time (Duffie and Beckman 2006), each of which can give a different perception of value. Here, we characterize the sensitivity of PV payback time to several system parameters using both the simple payback time definition (Kastovich et al. 1982; Perez et al. 2004; Black 2009) and the TNP payback definition (Nofuentes et al. 2002; Sidiras and Koukios 2005; Audenaert et al. 2010).

Although there are several other payback definitions, the relative performance of these two metrics illustrates how different definitions of payback can lead to a large range in payback times and sensitivities.

Figure 5 shows simple and TNP payback times for a range of non-price system characteristics. Based on the common definition of simple payback time (Table 1), the simple payback metric is insensitive to financing terms, discount rates, and electricity-rate increases since only the first year of revenue is considered in the formulation used. Simple payback times are mainly affected by increasing system revenue (electricity rates and capacity factors) or lowering system cost (Figure 2) and remain longer than 10 years for nearly all variations in non-price project parameters.

TNP payback frequently shows significantly lower payback times than simple payback and exhibits strong threshold behavior. For residential systems, these include small improvements in loan term, loan rate, capacity factor, and higher electricity rates. For commercial systems, these include shorter loan terms, higher loan rates, lower capacity factors, lower electricity rates, and lower tax rates. The threshold behavior for payback times is not unique to the TNP payback metric; several other payback definitions (Duffie and Beckman 2006) are likely to show similar threshold behavior for varying system price and non-price parameters.

Customers using payback metrics may be less likely to adopt PV than if they used other metrics to characterize PV value. Simple payback times are frequently on the order of decades, which may not entice customers to adopt, particularly residential customers who frequently internalize high discount rates when valuing energy saving investments (Wilson and Dowlatabadi 2007). Although the TNP payback metric frequently shows very short payback times, the large sensitivity to system assumptions, seen by the strong threshold behavior, may confuse potential customers as to the actual value of a PV system, which may reduce adoption because of customer aversion to uncertainty (Wilson and Dowlatabadi 2007). Also, PV customers using simple payback times to characterize PV value are insensitive to improving several non-price system parameters, and they may only be enticed to adopt PV if the variables they are sensitive to, system price and revenues, are significantly improved. This again represents a strong opportunity for the public sector to educate potential PV customers about the value of a PV investment as shown through other economic performance metrics.

Figure 6 shows annualized MBS for residential and commercial PV customers for a range of non-price system characteristics. MBS are shown for both customer-owned and third-party owned systems. MBS from third-party PV systems are typically higher, based on the different tax implications of system ownership, as discussed in Section 4.5 and Appendix A. Although all PV economic metrics would show different returns based on the ownership structure, we only highlight the differences for MBS because third-party PV companies frequently market products to customers based on MBS (NREL 2009; SolarCity 2011; SunRun 2011). However, several of the trends shown for MBS are similar across other metrics.

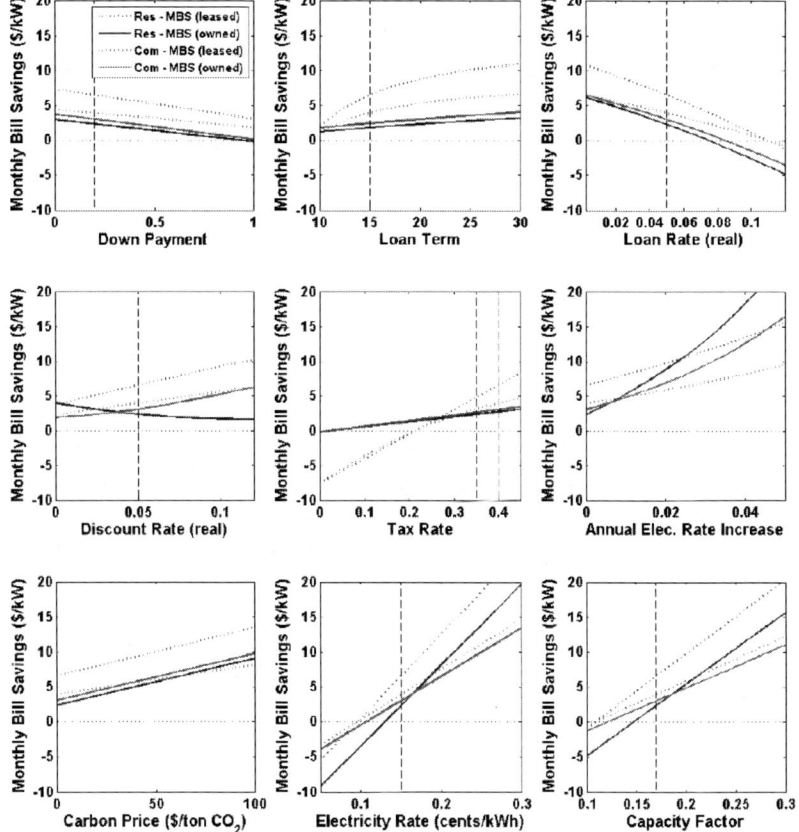

Note: Vertical dashed lines show reference PV assumptions, and horizontal dotted grey lines show the demarcation between a profitable and unprofitable investment.

Figure 6. Annualized MBS for $4,000/kW residential and commercial PV systems.

MBS shows positive savings for the reference parameters, and they generally increase by spreading system costs over more years, decreasing system costs, and increasing revenues. For the reference conditions, MBS ranges from $1.27–$5.46/kW-month for residential systems and $1.66–$3.28/kW-month for commercial systems. The savings shown in Figure 6 are generally higher for third-party owned systems than for customer-owned systems because of the different tax structures (see Appendix A). These savings are based on the reference PV cost of capital and financing assumptions, but actual third-party owned MBS offerings may be lower based on higher capital costs and overhead. The actual MBS (or monthly costs) seen by customers is based on the size of their PV system. For example, residential PV systems are typically about 5 kW, and commercial systems are around 100 kW, leading to annual bill savings that are on the order of $76–$330/yr for residential customers and $2,000–$3,900/yr for commercial customers.

The reference PV cost and performance parameters represent positive MBS, which may be more attractive to potential PV customers than framing the system in terms of a simple payback time on the order of tens of years, particularly for residential customers who typically internalize high discount rates when evaluating investment choices (Wilson and Dowlatabadi 2007). This suggests that third-party PV companies have a strong opportunity to attract customers by repackaging costs and revenues into a simple product that generates MBS (NREL 2009; Drury et al. 2011).

Figure 7 shows residential and commercial LCOEs for a range of non-price system characteristics. Since LCOE is a relative metric that must be compared to the value of electricity generated, we also show the assumed effective electricity rate ($0.15/kWh) on each figure. PV LCOEs that are lower than the effective electricity price may represent a profitable investment, and PV LCOEs that are higher may represent an unprofitable investment. LCOEs are unaffected by varying electricity rates, rate increases, and carbon prices, and these sensitivities are not shown in Figure 7. However, varying these parameters will modify electricity rates, and an increase for each parameter will increase the value of a PV investment.

PV LCOEs are lower than the assumed effective electricity rate for the reference assumptions and for several variations in system parameters, and these may represent profitable PV investment opportunities. Like other metrics, PV LCOEs improve by decreasing system costs, spreading costs over more years, and increasing revenues. However, the value of a PV investment is directly tied to the relationship between LCOEs and the value of PV electricity, which is based on the characteristics of the local electricity generation fleet, load profiles, fuel prices, transmission constraints, net-metering policy, and several other variables. PV LCOEs are also frequently compared to the LCOEs of other electricity-generating technologies with different generation profiles, which can lead to an incorrect estimation of value (Borenstein 2008). These, and other, factors need to be taken into account for LCOE to produce an accurate measure of PV value.

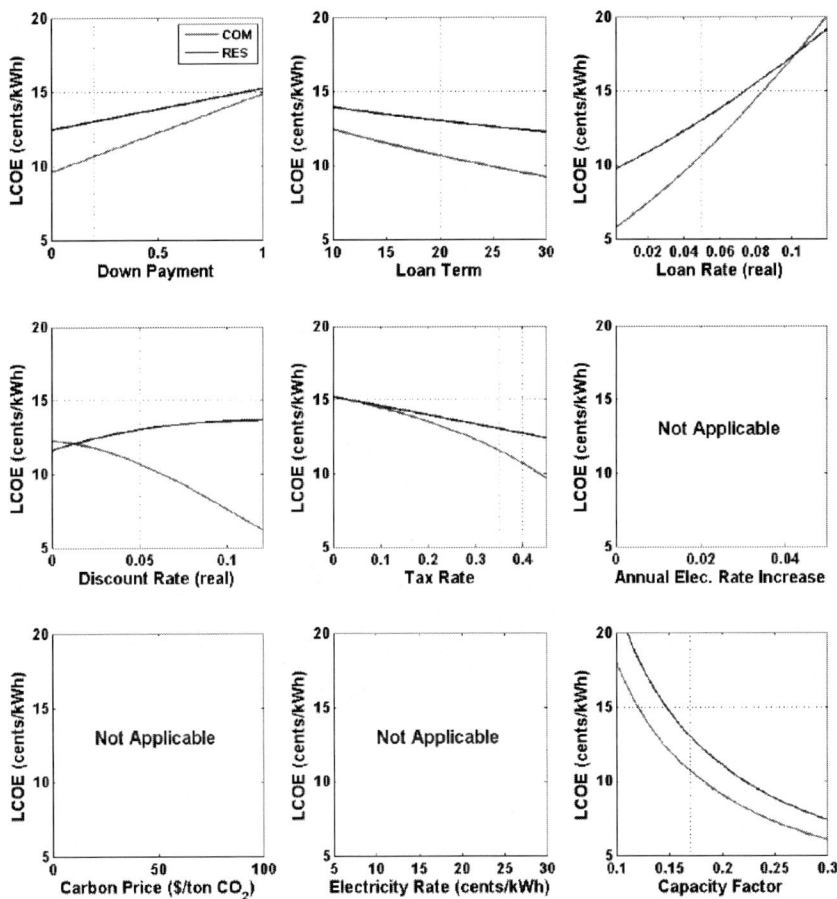

Note: The LCOEs shown here are in units of real dollars, not nominal, dollars.

Figure 7. LCOEs for $4,000/kW residential and commercial PV systems over a range of financing, performance, and market parameters.

8. RESULTS ACROSS PERFORMANCE METRICS

One challenge in comparing PV economic performance across several metrics is that each metric typically uses different units to characterize PV value (Table 1). These include years for payback times, annualized returns for IRR, total discounted returns for B/C ratio and PI, dollars for NPV and dollars per month for MBS, and the cost of generating electricity for LCOE. Although we can evaluate the relative sensitivity of each metric to a range of system parameters, it is challenging to compare these sensitivities between metrics. We address this here by defining a common unit of performance that allows us to compare relative sensitivities between metrics. We define this metric as the "Equivalent Change in PV Price," which characterizes the impact of varying non-price system parameters in units of an equivalent change in PV price that produces identical returns. We calculate the Equivalent Change in PV Price using four steps:

1) Calculate reference PV performance for each economic metric. This represents a $4,000/kW PV system and the reference financing, market, and performance parameters listed in Table 2.
2) Vary one of the non-price system parameters and recalculate PV performance for a $4,000/kW system. This generates a second return for each metric.
3) Using reference non-price system parameters solve for a new PV price that generates the second return (found in step 2).
4) Define the Equivalent Change in PV Price as the new PV system price (found in step 3), minus the reference system price.

For example, to calculate the Equivalent Change in PV Price when varying the loan rate from 5% to 3% (real) for a residential system using the PI metric, we: (1) calculate PI for a reference residential PV system and find PI = 11.2%; (2) vary the loan rate from 5% to 3% (real) and find that PI increases to 19.2%; (3) return to the reference non-price parameters, including a 5% (real) loan rate, and vary PV price until we solve for the PV price that gives us a PI of 19.2%; we find that a PV system price of $3,595/kW will generate a 19.2% PI for a system with a 5% (real) loan rate; and (4) define the Equivalent Change in PV Price as the new system price minus the reference system price ($4,000/kW) and find an Equivalent Change in PV Price of −$405/kW. Conceptually, this means that system PIs could be increased from 11.2% to 19.2% by decreasing the loan rate from 5% to 3% (real) or by decreasing the system price from $4,000/kW to $3,595/kW. In this way, the value of decreasing loan rates from 5% to 3% (real) is equal to reducing capital costs by $405/kW for the PI metric.

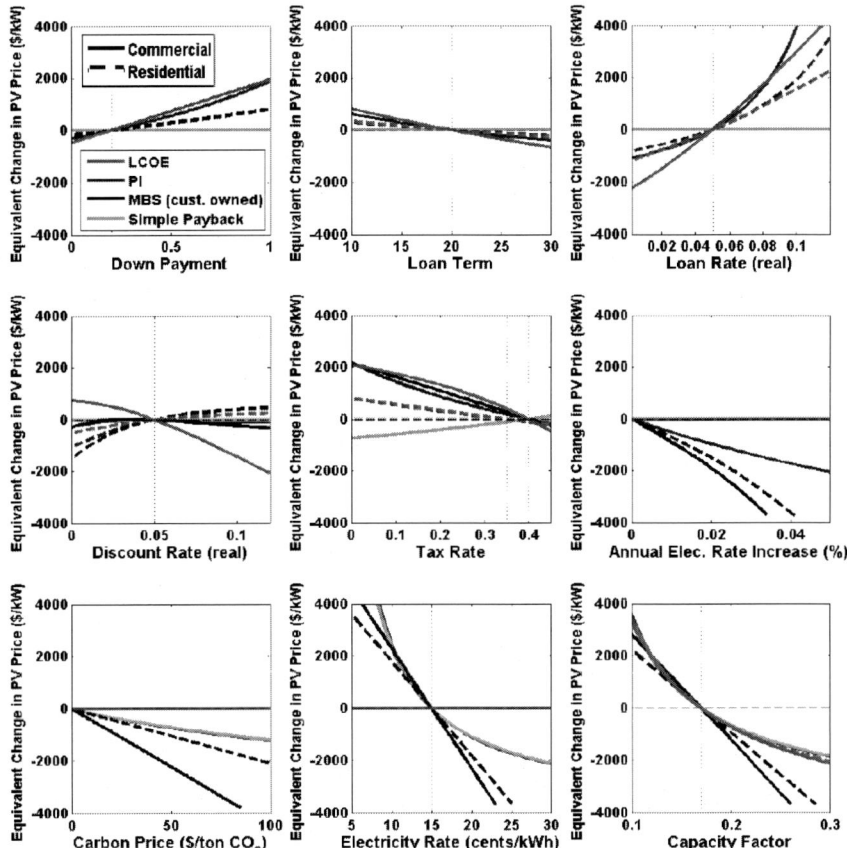

Note: Vertical dashed lines show the reference PV assumptions, which are assumed to be identical for residential and commercial systems except for the effective tax rate (35% for residential, 40% for commercial).

Figure 8. Equivalent Change in PV Price for the PI, LCOE, simple payback time (payback), and customer-owned MBS metrics.

We then use the same method for several economic performance metrics and find different Equivalent Changes in PV Price for each variation in non-price parameter. For example, using the LCOE metric, we find that reducing residential loan rates from 5% to 3% (real) leads to a higher Equivalent

Change in PV Price of −$533/kW. This suggests that the LCOE metric is more sensitive to the loan rate parameter than the PI metric for the reference residential system.

Figure 8 shows the Equivalent Change in PV Price for residential and commercial systems for several economic performance metrics and a range of non-price project parameters. Most metrics show similar trends (all positive or all negative slopes), although the strength of these trends varies. Metrics with steeper slopes are more sensitive to the non-price variable than metrics with flatter slopes. For example, commercial LCOEs show an Equivalent Change in PV Price of about $2,000/kW if the down payment is increased from 20% to 100%, whereas residential LCOEs show about a $1,000/kW Equivalent Change in PV Price for the same increase in down payment fraction. Different metrics can also show different trends, illustrated by a positive slope for one metric and a negative slope for another metric.

For example, the commercial LCOE shows a positive Equivalent Change in PV Price for a discount rate equal to zero, while all other metrics show a negative or zero Equivalent Change in PV Price.

Most metrics show different Equivalent Changes in PV Price to varying system parameters. These differences show how evolving market conditions, financing terms, or carbon policy could impact some market participants more strongly than others. For example, if loan rates (real) were to decrease broadly across all customer classes, PV MBS and LCOEs would improve more significantly than system PIs. This suggests that access to low-cost capital could preferentially stimulate market segments using MBS (likely residential) and LCOE (likely large-scale) relative to other market segments using PI (likely commercial rooftop).

Also, since MBS improves significantly while simple payback times remain static, access to low-cost capital could preferentially stimulate third-party owned PV adoption, which is typically marketed in terms of MBS, relative to customer-owned PV adoption where residents frequently think in terms of investment payback times. Similar trends are shown for several other system parameters.

The Equivalent Change in PV Price results shown in Figure 8 also quantify, in terms of dollars per kilowatt, the impact of varying several non-price system characteristics. These impacts can be compared to capacity-based or investment-based incentives, such the federal ITC or PV rebates offered by several states (DSIRE 2011), as additional methods for stimulating PV markets.

However, since the impacts of varying non-price project parameters are often inconsistent between different economic metrics, it is important to understand these sensitivities for designing effective policy.

9. POLICY IMPLICATIONS

Several types of incentives have been developed to spur renewable energy deployment worldwide. The majority of U.S. solar incentives are either capacity-based (reducing project costs by a fixed amount per unit of installed capacity) or investment-based (reducing project costs by a fixed fraction) (DSIRE 2011). These incentives are typically paid during the first year of ownership. In contrast, U.S. wind projects and European solar projects frequently receive production-based incentives, which are paid over several years based on the amount of electricity generated by a project. We find that the timing of project costs and revenues, including incentives, is a key driver for the economic returns generated by a PV investment (Sections 7–8). In general, an upfront incentive has more impact per dollar spent than an incentive spread over several years. However, the sensitivity to the timing of incentives is metric-dependent, and switching from one incentive type to another could preferentially stimulate individual market segments relative to others.

In addition to traditional incentives, we find that PV returns can be improved significantly by modifying several non-price project parameters. In particular, attractive financing terms can improve PV economics by spreading costs over more years (e.g., lower down payment fraction and longer loan term) and by decreasing investment costs (e.g., lower capital costs). We find that varying PV financing parameters can impact project economics by an amount that is equivalent to increasing or decreasing PV project costs by several dollars per watt (Figure 8). Providing renewable energy projects with access to long-term, low-cost financing may be a cost-effective method for increasing demand at lower costs than providing capacity-based, investment-based, or production-based incentives.

Since the perceived value of a PV investment can be shaped by the economic metric used to characterize performance, there is a strong potential for increasing PV demand by educating potential customers about the value of a PV investment as represented by several metrics. Behavioral economics suggests that customers respond strongly to how information is presented, which is called the "framing effect" (Wilson and Dowlatabadi 2007). For example, a residential customer who uses simple payback time to characterize the value of a PV investment may be far less likely to invest in PV than a similar residential customer characterizing the same investment in terms of MBS. The customer using simple payback time may require lower PV prices, either through direct incentives or PV price and performance improvements, before they are willing to adopt PV, whereas the customer using MBS may be enticed to adopt the given system. Improving the information available to customers could potentially reduce disparities in adoption trends both within and between market segments.

Lastly, new business models like third-party PV ownership can repackage PV costs and revenues into simple products like MBS, and allowing these businesses to enter the market could stimulate PV demand. For example, mean annualized MBS can be calculated for a customer-owned system; however, the actual costs and revenues generated by the PV system will vary significantly from the annualized MBS on a monthly basis. For customers to see a PV investment in terms of MBS may require a third-party company to repackage PV returns into a simple product. In this way, third-party business models can fundamentally reshape the perception of PV value. Several states have policies that limit third-party PV ownership (Kollins et al. 2010), and this represents a strong opportunity for the public sector to engage state and local officials to reduce the barriers to entry.

CONCLUSION AND FUTURE WORK

PV is adopted by several types of market participants, ranging from residential customers to utilities and large-scale developers. We find that the use of different economic performance metrics by each market participant can significantly shape the perceived value of a PV investment. This can lead to different prices for when a PV investment looks profitable or attractive and different sensitivities to non-price system parameters.

The upfront nature of U.S. PV incentives can lead to challenges in calculating and interpreting PV value using some economic performance metrics. For example, we find that the upfront nature of U.S. incentives make IRRs and MIRRs poor metrics for characterizing PV economics. U.S. incentives can make IRRs very sensitive to changing system price and non-price parameters and generally lead to an inflated perception of potential returns. The same incentives make MIRRs relatively insensitive to varying system price and non-price parameters. Also, the upfront nature of U.S.

incentives can lead to large differences in the economic returns calculated for U.S. systems relative to similar systems located in other countries and discussed in the international PV literature (e.g., Nofuentes et al. 2002; Talavera et al. 2007; Talavera et al. 2010; Audenaert et al. 2010).

That the perceived value of a PV investment can be significantly shaped by the choice of economic performance metric has important implications for policy design. For example, enabling access to long-term, low-cost capital could preferentially stimulate markets that use economic performance metrics that are sensitive to financing parameters (commercial and some residential), while having little or no impact on customers using metrics that are insensitive to financing terms (some residential). It is critical that policymakers understand these metric-dependent sensitivities to design effective policy.

This analysis suggests several areas for improving our understanding of how customers make adoption decisions and how policy affects these decisions. This represents an opportunity to learn from PV customers and to better understand their concerns and priorities when evaluating a potential PV investment. This also represents an opportunity to educate potential customers about the value of a PV investment as seen through several economic performance metrics and help them make more informed adoption decisions.

Improving our understanding of customer behavior can also be used to improve the representation of adoption behavior in PV market penetration models (EIA 2008a; EIA 2008b; Paidipati et al. 2008; Denholm et al. 2009; R.W. Beck 2009; Drury et al. 2010). These models frequently use one economic metric for each customer type, typically a payback time, to characterize adoption behavior. Modeled depictions of market evolution and the impacts of new policy are shaped by the sensitivities of one economic performance metric and do not capture the impacts of market participants using several metrics or evolving customer behavior. This represents a strong opportunity for improving the representation of PV value and adoption behavior in models, which can be used to better inform policy design.

REFERENCES

Audenaert, A.; De Boeck, L.; De Cleyn, S.; Lizin, S.; Adam, J-F. (2010). "An Economic Evaluation of Photovoltaic Grid Connected Systems (PVGCS) in Flanders for Companies: A Generic Model." *Renewable Energy* (35); pp. 2674–2682.

Barbose, G.; Darghouth, N.; Wiser, R. (2011). *Tracking the Sun IV: An Historical Summary of the Installed Cost of Photovoltaics in the United States from 1998–2010.* LBNL-5047E. Berkeley, CA: Lawrence Berkeley National Laboratory.

Black, A. (2009). "Economics of Solar Electric Systems for Consumers: Payback and Other Financial Tests." OnGrid Solar Energy Systems. http://www.ongrid.net/papers/PaybackOnSolarSERG.pdf. Accessed February 2011.

Bolinger, M. (2009). *Financing Non-Residential Photovoltaics Projects: Options and Implications.* LBNL-1410E. Berkeley, CA: Lawrence Berkeley National Lab.

Borenstein, S. (2008). *The Market Value and Cost of Solar Photovoltaic Electricity Production.* CSEM WP 176. University of California Energy Institute, Center for the Study of Energy Markets, Berkeley, CA.

California Energy Commission (CEC). (June 2007). "Comparative Costs of California Central Station Electricity Generation Technologies." CEC-200-2007-011-SD.

Chabot, B. (1998). "From Cost to Prices: Economic Analysis of Photovoltaic Energy and Services." *Progress in Photovoltaics: Research and Applications,* (6); pp. 55–68.

Database of State Incentives for Renewables and Efficiency (DSIRE). (2011). http://www.dsireusa.org, Accessed March 2011.

Denholm, P.; Drury, E.; Margolis, R. (2009). *Solar Deployment System (SolarDS) Model: Documentation and Sample Results.* NREL/TP-6A2-45832. Golden, CO: National Renewable Energy Laboratory.

Denholm, P.; Margolis, R.; Ong, S.; Roberts, B. (2009). *Break-Even Cost for Residential Photovoltaics in the United States: Key Drivers and Sensitivities.* NREL/TP-6A2-46909. Golden, CO: National Renewable Energy Laboratory.

Drury, E.; Denholm, P.; Margolis, R. (2010). *Modeling the U.S. Rooftop Photovoltaics Market.* NREL/CP-6A2-47823.Golden, CO: National Renewable Energy Laboratory.

Drury, E.; Miller, M.; Heimiller, D.; Graziano, D.; Macal, C.; Ozik, J.; Perry, T. "The Transformation of southern California's Residential Photovoltaics Market through Third-Party Ownership", *submitted to Energy Policy, July 2011.*

Duffie, J.A.; Beckman, W.A. (2006). *Solar Engineering of Thermal Processes.* 3rd ed. Hoboken, NJ: John Wiley & Sons, Inc., 908 pp.

Energy Information Administration (EIA). (2008a). "Commercial Sector Demand Module of the National Energy Modeling System: Model Documentation 2008." DOE/EIA-M066(2008). Washington, DC: Energy Information Administration.EIA. (2008b). "Residential Sector Demand Module of the National Energy Modeling System." DOE/EIA-M067(2008). Washington, DC: Energy Information Administration.

EIA. (2011a). "Annual Energy Outlook 2011 with Projections to 2035." DOE/EIA-0383(2011). http://www.eia.doe.gov/forecasts/aeo/. Accessed June 2011.

EIA. (2011b). "Electric Power Monthly, Table 5.6.A., Average Retail Price of Electricity to Ultimate Customers by End-Use Sector, by State, June 2011 and 2010." http://www.eia.gov/electricity. Accessed September 2011.

Harper, J.P.; Karcher, M.D.; Bolinger, M. (September 2007). *Wind Project Financing Structures: A Review and Comparative Analysis.* LBNL-63434. Berkeley, CA: Lawrence Berkeley National Laboratory.

Internal Revenue Services (IRS). (2010). "Home Mortgage Interest Deduction." Publication 936. Washington, DC: U.S. Department of the Treasury. http://www.irs.gov/pub/irs-pdf/p936.pdf. Accessed August 2011.

Kastovich, J.; Lawrence, R.; Hoffmann, R.; Pavlak, C. (1982). *Advanced Electric Heat Pump Market and Business Analysis.* ORNL/Sib/79-2471/1. Prepared under subcontract to ORNL by Westinghouse Electric Corp. Oak Ridge, TN: Oak Ridge National Laboratory.

Kollins, K.; Speer, B.; Cory, K. (2010). *Solar PV Project Financing: Regulatory and Legislative Challenges for Third-Party PPA System Owners.* TP-6A2-46723. Golden, CO: National Renewable Energy Laboratory.

McKinsey & Co. (20 October 2004). "Internal Rate of Return: A Cautionary Tale," *The McKinsey Quarterly,*

Nofuentes, G.; Aguilera, J.; Munoz, F.J. (2002). "Tools for the Profitability Analysis of Grid-Connected Photovoltaics." *Progress in Photovoltaics: Research and Applications* (10); pp. 555–570.

Network for New Energy Choices (NNEC). (2010). *Freeing the Grid,* 4th ed. http://www.newenergychoices.org/uploads/FreeingTheGrid2010.pdf. Accessed July 2011.

National Renewable Energy Laboratory (NREL). (2009). *Solar Leasing for Residential Photovoltaic Systems.* NREL/FS-6A2-43572. Golden, CO: National Renewable Energy Laboratory. http://www.nrel.gov/docs/fy09osti/43572.pdf. Accessed February 2011.

Paidipati, J.; Frantzis, L.; Sawyer, H.; Kurrasch, A. (February 2008). *Rooftop Photovoltaics Market Penetration Scenarios.* NREL/SR-581-42306. Golden, CO: National Renewable Energy Laboratory.

Perez, R.; Burtis, L.; Hoff, T.; Swanson, S.; Herig, C. (2004). "Quantifying Residential PV Economics in the U.S. – Payback vs. Cash Flow Determination of Fair Energy Value." *Solar Energy* (77); pp. 363–366.

R.W. Beck. (January 2009). *Distributed Renewable Energy Operating Impacts and Valuation Study.* Prepared for Arizona Public Service by R.W. Beck, Inc.

Short, W.; Packey, D.; Holt, T. (1995). *A Manual for the Economic Evaluation of Energy Effiency and Renewable Energy Technologies.* NREL/TP-462-5173. Golden, CO: National Renewable Energy Lab.

Sidiras, D.K.; Koukios, E.G. (2005). "The Effect of Payback Time on Solar Hot Water Systems Diffusion: The Case of Greece." *Energy Conversion and Management.* (46); pp. 269–280.

Solar Energy Industries Association and Greentech Media Research (SEIA-GTM). (2011). "U.S. Solar Market Insight, 1st Quarter 2011." http://www.greentechmedia.com/research/solarinsight. Accessed June 2011.

Solar Energy Industries Association (SEIA). (2009). "Guide to Federal Tax Incentives for Solar Energy," Version 4.1. Washington, DC: Solar Energy Industries Association.

SolarCity. (2011). http://www.solarcity.com. Accessed July 2011.

Stermole, F.J.; Stermole, J.M. (2009). "Economic Evaluation and Investment Decision Methods." Lakewood, CO: Investment Evaluations Corporation.

Sun Light & Power. (2011). http://www.sunlightandpower.com/. Accessed July 2011.

Sunpower Corp. (August 2008). "The Drivers of the Levelized Cost of Electricity for Utility-Scale Photovoltaics." San Jose, CA: Sunpower Corporation. http://nl.sunpowercorp.be/downloads/SunPower levelized cost of electricity.pdf. Accessed July 2011.

SunRun. (2011). http://www.sunrunhome.com. Accessed July 2011.

System Advisor Model (SAM). (2011). SAM version 2010.11.9, National Renewable Energy Laboratory. https://www.nrel.gov/analysis/sam/. Accessed March 2011.

Talavera, D.L.; Nofuentes, G.; Aguilera, J.; Fuentes, M. (2007). "Tables for the Estimation of the Internal Rate of Return of Photovoltaic Grid-Connected Systems." *Renewable and Sustainable Energy Reviews* (11); pp. 447–466.

Talavera, D.L.; Nofuentes, G.; Aguilera, J. (2010). "The Internal Rate of Return of Photovoltaic Grid-Connected Systems: A Comprehensive Sensitivity." *Renewable Energy* (35); pp. 101–111.

Wilson, C.; Dowlatabadi, H. (2007). "Models of Decision Making and Residential Energy Use." *Annual Review of Environment and Resources* (32); pp. 169–203.

APPENDIX A. TAX IMPLICATIONS FOR THIRD-PARTY PV OWNERSHIP

Monthly bill savings (MBS) are frequently used by third-party owned PV companies and other PV retailers to characterize the value of a PV system to potential customers (NREL 2009; SolarCity 2011; SunRun 2011). In this analysis, we define the annualized MBS of a PV system based on the difference between the PV LCOE and the effective electricity rate, multiplied by the amount of electricity generated by the PV system, as shown in Equation A.1:

$$MBS = \frac{1}{LeaseTerm * 12} \sum_{t=1}^{LeaseTerm} \frac{PV\ Generation_t * (Electricity\ Rate_t - LCOE)}{(1+d)^t} \quad (A.1)$$

Here, *PV Generation$_t$* is the annual amount of electricity generated by a PV system in a given year, *Electricity Rate$_t$* is the effective electricity rate that represents the annualized value of hourly PV generation for a given year, and *PV LCOE* is calculated for a standard financed system (Table 1).

PV MBS can vary significantly for different ownership models. If the PV system is owned by a residential customer, the *Electricity Rate$_t$* is based on full retail electricity prices. If the PV system is owned by a commercial customer, the *Electricity Rate$_t$* is based on tax-deductible electricity prices, and the PV *LCOE* accounts for MACRS depreciation. If the PV system is owned by a third-party PV company, we assume that the PV *LCOE* accounts for MACRS depreciation, for both residential and commercial site hosts, and a higher tax rate than residential customers.

Different ownership structures could potentially lead to higher MBS offerings from third-party owned systems, based on different tax burdens for each ownership structure. For residential systems, access to MACRS and higher commercial tax rates could produce higher MBS offerings for systems with identical prices, electricity rates, and financing terms.[20] The difference for commercial systems is not likely to be as great because they already depreciate PV assets based on MACRS. However, third-party MBS offerings could potentially be higher because of differences in taxing energy costs. For a customer-owned system, the effective electricity price is the tax-deductible retail electricity rate representing the value of avoided electricity use, and PV LCOEs have to be lower than this to produce a bill savings. For third-party systems, the PV LCOE only has to be lower than the retail electricity rate to produce bill savings, as shown in Equations A.2–A.4:

$$MBS_{Customer\ Owned} = \frac{1}{LeaseTerm*12} \sum_{t=1}^{LeaseTerm} \frac{PV\ Generation_t * (Electricity\ Rate_t * (1 - Tax\ Rate) - LCOE)}{(1+d)^t} \quad (A.2)$$

$$MBS_{Third-party} = \frac{1}{LeaseTerm*12} \sum_{t=1}^{LeaseTerm} \frac{PV\ Generation_t * Electricity\ Rate_t * (1 - Tax\ Rate) - Lease/PPA\ Cost_t * (1 - Tax\ Rate)}{(1+d)^t} \quad (A.3)$$

$$MBS_{Third-party} = \frac{1}{LeaseTerm*12} \sum_{t=1}^{LeaseTerm} \frac{PV\ Generation_t * (Electricity\ Rate_t - LCOE) * (1 - Tax\ Rate)}{(1+d)^t} \quad (A.4)$$

Here, all variables have the same definitions as those in Equation A.1, and *Lease/PPA Cost$_t$* represents either the annual cost for leasing PV equipment or the annual cost for purchasing PV electricity through a PPA. Equation A.2 represents the annualized MBS for a customer-owned commercial system, and Equation A.3 represents the MBS for a third-party owned PV system that is either leased to a commercial customer or the electricity sold to a commercial customer through a PPA.

If the system *Lease/PPA Costt* can be approximated as the LCOE of the system times the amount of energy generated by the system, Equation A.3 simplifies to Equation A.4. Equations A.2 and A.4 show the difference in customer-owned and third-party owned commercial systems, where the LCOE of a customer-owned system needs to be less than the tax-deductible electricity rate to generate bill savings while the LCOE of the third-party owned system may only need to be lower than the effective electricity rate to generate savings. Equations A.3 and A.4 do not factor in how the tax burden of the third-party company impacts potential MBS offerings. If all third-party PV revenues were taxed, and the third-party company had the same marginal tax rate as the commercial client, Equation A.4 would trend toward equation A.2, and there would be no tax benefit for third-party ownership. However, third-party PV companies pay taxes on net revenues (where MBS would trend toward equation A.4), not total revenues (where MBS would trend toward

A.2). This suggests that third-party PV companies may have a competitive advantage relative to residential and commercial customers buying their own systems, based on tax structure.

APPENDIX B. THE CHALLENGE OF INTERPRETING INTERNAL RATES OF RETURN FOR U.S. PV SYSTEMS

Several types of investors use the IRR to characterize and rank investment returns for a range of investment opportunities. This is particularly true for wind developers and tax investors, where IRR-based hurdle rates[21] are often used to rank potential wind projects (Harper et al. 2007). IRR is also frequently used to evaluate PV economic performance, particularly in European markets (Nofuentes et al. 2002; Talavera et al. 2007; Talavera et al. 2010; Audenaert et al. 2010). Although IRR can be useful for characterizing the value of projects that receive no incentives, or production-based incentives, it frequently shows threshold effects (Talavera et al. 2010) can have multiple positive real solutions (Stermole and Stermole 2009) and can inflate the perceived value of an investment if the IRR is significantly higher than the opportunity cost of capital (McKinsey & Co. 2004). The challenges with IRR are even greater for U.S. PV systems because the combination of the 30% federal ITC, along with MACRS for commercial systems, exacerbates threshold behavior and inflates the perception of value. Because of these issues, we find that IRR is a poor metric for characterizing the value of U.S. PV systems.

There are three main challenges for interpreting PV IRR: (1) IRR frequently has multiple positive, real solutions, (2) IRR is subject to strong threshold behavior, and (3) IRR frequently inflates the perceived value of a PV investment. PV IRR frequently has multiple solutions because solving for the IRR of a PV system with a 20-year lifetime entails finding the solution to a 20th-order polynomial, which can have up to 20 solutions. Often, there is only one IRR solution that is both positive and real.[22] However, it is not uncommon for PV cash flows to generate several solutions that are both positive and real, particularly for financed PV systems. This is primarily based on the upfront nature of PV costs (down payment) and U.S. incentives (federal ITC plus MACRS depreciation for commercial customers), where PV cash flows can transition from negative (down payment), to positive (incentives), to negative (if PV revenues are insufficient to fully offset loan payments and O&M costs), to positive (PV revenues after the loan term) (see Figure 1).

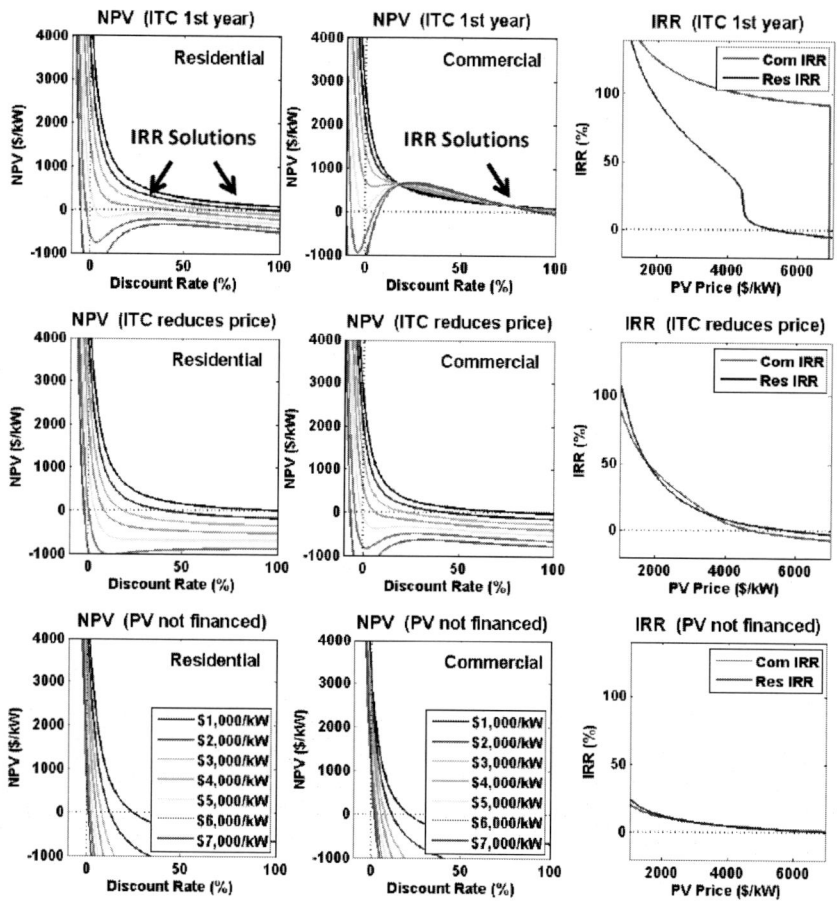

Note: IRR solutions represent the discount rates where the NPV(s) of a PV system equals zero.

Figure B.1. The relationship between residential and commercial NPV to discount rates for several PV prices and the corresponding relationship between residential and commercial IRRs and PV prices for three financing and incentive structures.

Figure B.1 shows the relationship between the system NPV and discount rate, where IRR solutions represent the discount rates in which NPV equals zero (Table 1). Also shown are system IRRs for a range of PV prices, which illustrate some of the challenges in interpreting IRR as a measure of PV investment value.

Figure B.1 shows that several systems have only one real IRR solution and an NPV-to-discountrate relationship that suggests the IRR solution is meaningful and representative of investment returns. For example, unfinanced residential and commercial PV systems show positive NPVs for discount rates equal to zero, a steep decline in NPVs with increasing discount rates, and a single IRR solution that gives a meaningful indication of investment returns.

However, the combination of system financing and incentives frequently complicates the interpretation of a project IRR. For example, a $6,000/kW financed commercial PV system that takes the federal ITC in the first year of ownership has a negative NPV for a discount rate of 0%, which transitions to positive NPVs for discount rates higher than 5% and then transitions to negative NPVs for discount rates above about 100% (Figure B.1). The negative NPV for a discount rate of 0% shows that the undiscounted PV revenues are less than the undiscounted system costs. However, NPVs become positive for discount rates above 5% because the discounting is sufficient to decrease the relative importance of the years with negative cash flows after MACRS expires (years 7–20 for the $6,000/kW system) and increase the relative importance of years with positive cash flows from the federal ITC and MACRS (years 1–7). NPVs become negative again for discount rates greater than 100%, which increases the relative importance of the loan down payment in the first year and decreases the importance of the positive cash flows from the federal ITC and MACRS payments over the next six years. The issue of multiple IRR solutions is resolved for: (1) commercial systems that account for the federal ITC by reducing the system price rather than as a positive revenue source after the first year of ownership; (2) residential systems that do not receive MACRS depreciation; and (3) residential and commercial systems that are not financed.

Another challenge in interpreting investment returns for systems with high IRR solutions is that the timing of PV cash flows frequently leads to very high IRR solutions that show small but positive NPVs. For example, a $4,000/kW residential PV system that takes the federal ITC in the first year of ownership shows an IRR of 42%, but system NPVs are less than $100/kW (equivalent to PI < 2.5%) for discount rates above 21%. The very slow decrease in NPV with increasing discount rates is primarily caused by the upfront nature of PV costs (down payment) and incentives (ITC and MACRS depreciation for commercial systems), where very high discount rates are required to affect the relative balance between the loan down payment and the federal ITC.

Several previous studies have focused on the utility of IRR to characterize PV value (e.g., Talavera et al. 2010; Talavera et al. 2007). However, these studies did not analyze the economics of PV systems receiving large upfront capacity-based or investment-based incentives, such as those in the United States. We find that IRR values are frequently misleading for U.S. systems and are poor measures of PV value.

[1] State and local rebates frequently increase a customer's federal tax basis, although there are some exceptions based on ownership or project structure (SEIA 2009). We assume that the effective PV price includes all tax payments (e.g., sales tax and federal tax on state and local rebates) and is not subject to further taxation.

[2] Nominal interest rates or dollars account for the effects of inflation, and the value of one nominal dollar changes from year to year. Real interest rates or dollars are adjusted to exclude inflation, and the value of one real dollar does not change over time. Nominal dollars are frequently converted to real dollars using the consumer price index (CPI) or gross domestic product (GDP), depending on application.

[3] PV economics are characterized in this analysis assuming dedicated PV financing and a discount rate that is equal to the loan rate. Large-system developers and third-party PV companies typically use more complex project financing, which include a range of tax equity, equity, and debt investors (e.g., Harper et al. 2007). The impacts of more complex project financing, and potentially higher capital costs, are not characterized in this analysis but will be the focus of subsequent work.

[4] For example, the interest rate for a residential home equity loan may be much lower than the interest rate for commercial debt financing because the capital accrued in the house can be used to offset investment risk.

[5] Several other metrics are similar to PI but scale NPV by different costs, including the maximum capital exposure (maximum capital outflow before positive system revenues reduce cost outlays) (Stermole and Stermole 2009) or life-cycle cost (Nofuentes et al. 2002).

[6] The opportunity cost of capital is the rate of return that could be realized on alternative investments of equivalent risk (Stermole and Stermole 2009).

[7] A hurdle rate represents the minimum rate of return that a company or project manager is willing to accept before developing a project.

[8] For example, in the first quarter of 2011 PV prices varied from $3.85/W for utility-scale systems, to $5.35/W for commercial systems, to $6.41/W for residential systems (SEIA-GTM 2011). However, electricity prices are frequently twice as high for residential retail markets as they are for wholesale electricity markets (EIA 2011a).

[9] The interest paid on a residential home mortgage, or a home equity loan up to $100,000, is tax deductible (IRS 2010).

[10] Net metering is a market mechanism that sets the value of PV generation that exceeds electricity use over a given amount of time. In areas with full net metering, excess PV electricity is purchased by local utilities at retail electricity rates. Other areas have partial net-metering policies in which excess PV generation is valued at prices that are similar to wholesale electricity rates and are roughly based on the value of offsetting fossil fuel use. Other areas have no net-metering policy and excess PV generation is not valued.

[11] Here and elsewhere, all PV costs, revenues, and market price projections are given in units of 2010 U.S. dollars.

[12] Although the application of PV incentives varies by state, the general trend is that state and local incentives are taken first, and then the 30% federal ITC is taken from the remaining system price. In this way, state and local incentives reduce the federal incentive.

[13] PV generation offsets after-tax energy costs for residential systems. However, commercial PV generation offsets tax-deductible energy costs, and commercial electricity rates are scaled by (1 − effective commercial tax rate) to generate after-tax returns (Figure 1).

[14] The reinvestment rate represents a company's opportunity cost of capital, or the returns they could expect on invested capital. For a PV system to be a profitable investment, relative to other investment opportunities, the PV MIRR must exceed the assumed reinvestment rate.

[15] We do not include NPV because all of the important trends and sensitivities are shown by the PI metric since it is equal to NPV normalized by the undiscounted initial investment price. NPV was shown in Figure 2 because the investment price was a variable, and the trends for NPV and PI are not identical.

[16] Here and elsewhere, the reference PV price includes all state and local incentives but not the federal ITC. PV price represents 2010 U.S. dollars.

[17] Decreasing the loan rate has two effects: (1) it reduces the cost of borrowing money, and by extension, the cost of the PV system, and (2) it can introduce a time value for money because the discount rate is held fixed at the reference loan rate. If the loan rate is lower than the discount rate, the discounted sum of loan payments is less than the initial price of the PV system, and the converse is true for loan rates that are higher than the discount rate.

[18] Carbon policy would increase the cost of electricity generated by burning fossil fuels. For example, coal-generated electricity produces about 1 kg CO_2/kWh, and natural gas generators produce about 0.45 kg CO_2/kWh (mean emissions from the combination of combustion turbine and combined cycle gas generation) (EIA 2011a). A carbon price of $20 per metric ton of CO_2 (spot price of carbon in the EU Emissions Trading System for January 2011) would add about $0.02/kWh to coal-generated electricity and about $0.01/kWh to electricity generated by natural gas.

[19] The B/C ratio differs from the (1 + PI) metric in that PI is calculated by dividing the NPV by the initial system price (Table 1), whereas the B/C ratio is calculated by dividing the discounted revenues by the discounted system cost. The B/C ratio frequently equals (1 + PI) when the discount rate is close to the loan rate, at a value where the discounted loan payments and O&M costs equal the initial system price. However, the discounted loan payments are higher than the initial system price if the discount rate is lower than the loan rate, and the converse is true if the discount rate is higher than the loan rate. This results in higher (1 + PI) returns for discount rates that are lower than the loan rate and higher B/C ratios for discount rates that are higher than the loan rate.

[20] This does not include the additional costs (financing costs, operating costs, additional overhead and margins) and benefits (economies of scale reached though high-volume installation) of third-party ownership, which will partially cancel.

[21] A hurdle rate represents the minimum rate of return that a company or project manager is willing to accept before developing a project.

[22] Several polynomial solutions will have non-zero imaginary components.

In: Solar Photovoltaics
Editors: G. Marks and C. J. Unger

ISBN: 978-1-62257-651-7
© 2013 Nova Science Publishers, Inc.

Chapter 2

BUILDING-INTEGRATED PHOTOVOLTAICS (BIPV) IN THE RESIDENTIAL SECTOR: AN ANALYSIS OF INSTALLED ROOFTOP SYSTEM PRICES[*]

National Renewable Energy Laboratory

LIST OF ACRONYMS

a-Si	Amorphous silicon
BAPV	Building-applied photovoltaics
BIPV	Building-integrated photovoltaics
c-Si	Crystalline silicon
CF	Capacity factor
CIGS	Cadmium indium gallium diselenide
DOE	U.S. Department of Energy
GW	Gigawatt (1 billion watts)
IEA	International Energy Agency
LCOE	Levelized cost of energy
Pmpp	Power maximum power point
PV	Photovoltaics

---

[*] This is an edited, reformatted and augmented version of National Renewable Energy Laboratory Technical Report, Publication No. NREL/TP-6A20-53103, dated November 2011.

SAM System Advisor Model
WACC Weighted average cost of capital
WVTR Water vapor transmission rates
$/W$_p$ DC U.S. (2010) dollars per peak watt of DC PV capacity

EXECUTIVE SUMMARY

For more than 30 years, there have been strong efforts to accelerate the deployment of solar-electric systems by developing photovoltaic (PV) products that are fully integrated with building materials. Despite these efforts and high stakeholder interest in building-integrated PV (BIPV), the deployment of PV systems that are partially or fully integrated with building materials is low compared with rack-mounted PV systems, accounting for about 1% of the installed capacity of distributed PV systems worldwide by the end of 2009. In this report, we examine the cost drivers and performance considerations related to BIPV for residential rooftops. We also briefly review the history of BIPV product development and examine market dynamics that have affected commercialization and deployment.

As with many renewable energy technologies, system prices—in terms of dollars per installed watt of direct-current peak power capacity ($/W$_p$ DC)— have a significant effect on PV deployment. In general, the installed prices of BIPV systems are higher than PV system prices, but the cause of these price premiums—higher costs, higher margins, or other considerations— and the potential for price reductions remain uncertain. Using a bottom-up analysis of components and installation labor costs, we explore the cost trade-offs that affect the prices of residential rooftop BIPV systems. We compare the prices of three hypothetical BIPV systems with the price of a rack-mounted crystalline silicon (c-Si) PV system, the "PV Reference Case," which is the most commonly installed residential system technology. One of the BIPV cases is a derivative of the c-Si PV case ("BIPV Derivative Case"), and the other two BIPV cases are based on an analysis of thin-film technologies (Table ES-1). In today's solar market, few BIPV products are fully integrated with building materials as envisioned in these BIPV cases; therefore, the cases should be seen as near-term possibilities. In contrast, the PV Reference Case represents a 2010 benchmark system price from an NREL study that uses the same methodology to assess objective system prices (Goodrich et al. 2011). Comparing the hypothetical near-term BIPV cases with the 2010 PV benchmark does not account for the continued advancements and cost reductions in rack-mounted PV systems. Thus, the potential cost advantages we have identified for BIPV installations are likely to change. Additionally, our analysis assumes that economies of scale and installer experience are equivalent for the PV Reference Case and the BIPV cases.

Table ES-1. Summary of Cases Used to Analyze Residential Rooftop PV System Prices

Scenario	Technology	Form	Efficiency	Module Area (m^2)
PV Reference Case	c-Si	Rigid	14.5%	1.28
BIPV Derivative Case	c-Si	Rigid	13.8%	0.58
BIPV Thin-film Case 1	CIGS	Rigid	11.2%	0.58
BIPV Thin-film Case 2	a-Si	Flexible	5.8%	0.58

a-Si—amorphous silicon; CIGS—Cu(In,Ga)Se$_2$; c-Si—crystalline silicon.

A summary of the analysis of PV and BIPV systems prices is shown below in Figure ES-1. The listed "effective prices" account for cost offsets due to an assumption that the BIPV cases replace traditional building materials; in this example, they replace asphalt shingles. Our findings suggest that BIPV has the potential to achieve system prices that are about 10% lower than rack-mounted PV system prices (i.e., the BIPV Derivative Case). The bulk of the BIPV cases' potential savings stem from eliminating the cost of module-mounting hardware—which rack-mounted PV systems need but BIPV systems do not—and from offsetting the cost of traditional building materials. BIPV labor savings result from the elimination of mounting hardware and our assumption of lower-cost roofing contractors in place of electricians. Some installation labor costs increase, however, due to the increased time that is required to install a greater number of smaller BIPV modules for a given area (i.e., more total electrical interconnections and wiring). Module costs and efficiencies are key factors that contribute to overall system prices across all of the cases, and we assume that the BIPV cases have lower efficiencies.

This report shows the potential for BIPV to achieve lower installed system prices than rack-mounted PV, but BIPV systems are likely to experience reduced performance (i.e., electricity generation) in comparison with PV systems. Unlike traditional PV systems that commonly include air spaces between the module and roof deck, BIPV systems are mounted directly on building surfaces, and this results in higher average operating temperatures in most environments. Resulting performance losses could affect the economic viability of BIPV projects. We assess comparative project economics by analyzing the unsubsidized levelized cost of energy (LCOE) for each case. Figure ES-2 summarizes the results of this LCOE analysis. The relative range of LCOE values differs from the relative range of installed system prices owing to differences in module efficiencies, degradation rates, and temperature coefficients.

These results show that c-Si BIPV shingles might achieve a lower LCOE than rack-mounted c-Si PV if installed system price advantages are fairly significant (i.e., greater than 5%). In cases where estimated BIPV cost advantages are small, expected performance losses may result in higher LCOE values, as shown in BIPV Thin-film Cases 1 and 2.

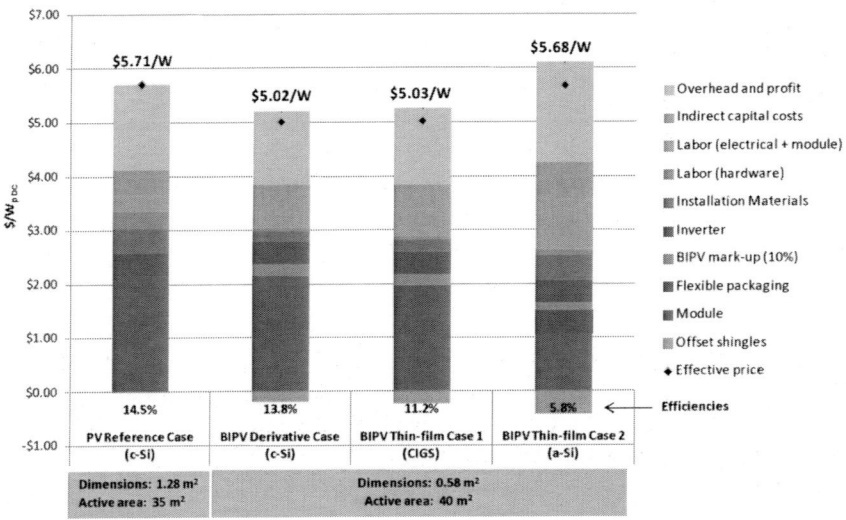

Note: Listed BIPV prices include building-material cost offsets (shown as negative bars).

Figure ES-1. Comparison of residential rooftop prices for a rack-mounted the PV Reference Case and three BIPV cases.

Overall, findings in this report support the notion that BIPV prices could be lower than residential PV system prices, yet past market experiences suggest that realizing these cost-reductions can be very challenging. To capitalize on the opportunities to reduce residential solar system prices and attract new consumers with aesthetically pleasing designs, BIPV faces more complex product-development issues and market-adoption dynamics than rack-mounted PV. We briefly address these less-quantifiable issues. An evaluation of specific commercial products goes beyond the scope of this report.

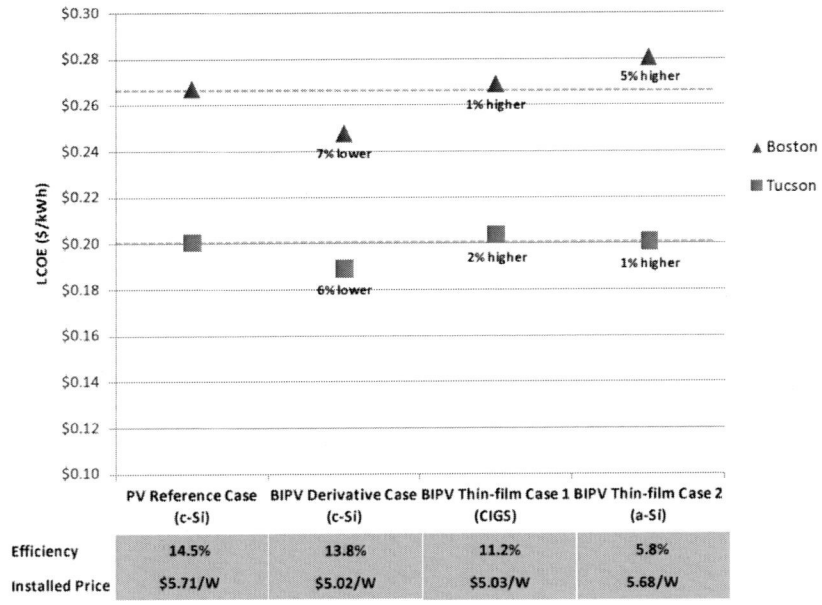

Note: Listed percentages illustrate LCOE differences relative to the PV Reference Case.[1] The LCOE calculations are based on consistent system price and financing structure assumptions for both locations—Boston and Tucson—but account for differences in estimated system prices, efficiencies, temperature coefficients, and degradation rates. All systems are south-facing and tilted at 25 degrees.

Figure ES-2. Unsubsidized U.S. residential rooftop LCOE values for the BIPV shingle cases compared with the PV Reference Case.

1. INTRODUCTION

Installations of solar photovoltaic (PV) technologies on building rooftops are common in some parts of the world. The vast majority of these systems are composed of modules that are mounted off the surfaces of roofs using different types of racking hardware. System designs are most influenced by PV performance considerations, and aesthetics are often secondary. But growing consumer interest in distributed PV technologies and industry competition to reduce installation costs are stimulating the development of multifunctional

PV products that are integrated with building materials. This emerging solar market segment, known as building-integrated PV (BIPV), continues to attract the attention of many stakeholders, as evidenced by the mention of a rooftop solar shingle product in the President's 2011 State of the Union Address (White House 2011).[2] BIPV offers a number of potential benefits, and there have been efforts to develop cost-competitive products for more than 30 years. The deployment of BIPV systems, however, remains low compared to traditional PV systems. In this report, we examine the status of BIPV, with a focus on residential rooftop systems, and explore key opportunities and challenges in the marketplace.

A continuum of PV system designs exists with various levels of integration with building materials and architectural features (Figure 1); there is no consensus definition of BIPV. Many stakeholders describe BIPV as a multifunctional product—one that acts as both a building material and a device that generates electricity (e.g., a solar shingle). Incentive programs and market reports, however, sometimes include *partially* integrated PV systems—those that blend with the designs of building materials but are not multifunctional—in their descriptions of BIPV. In Europe, for instance, the rules to qualify for BIPV-specific incentives are sometimes vague and include semi-integrated PV products (PV News 2010).[3] In many cases, semi-integrated products are a combination of PV products and traditional buildings materials (EPIA 2010). These combined products do not replace traditional building materials, and some stakeholders have described them as building-applied PV (BAPV).[4] Photon International describes BIPV modules as products that are "specifically constructed for building integration," and, in their recent survey of more than 5,000 commercially available modules, less than 5% were listed as BIPV.[5] Photon adds, however, that standard modules can also be integrated into buildings using certain mounting systems, implying that semi-integrated systems can also be described as BIPV (Photon International 2011). Regardless of the specific definitions of BIPV, it is clear that there is a continuum of integration with building materials among a class of PV products suited for rooftop and façade applications.

For this report, we consider BIPV to be a multifunctional product (not a combination of independent products) that generates electricity and replaces traditional building materials by serving as a significant weather barrier on residential building surfaces.[6] In other words, if the hypothetical BIPV cases we outline below were removed from rooftops, then repairs (e.g., waterproofing) would be required to ensure that buildings are protected from the environment. We call traditional, non-BIPV systems "rack-mounted PV";

these systems are intended to generate electricity only, are mounted on racks, and do not replace the function of building materials. The two photographs on the left in Figure 1 show examples of rack-mounted PV.

Least integrated **More integrated** **Fully integrated**
(Open rack-mounted PV) (Close roof rack-mounted PV) (Direct-mounted BIPV, multifunctional)

Source: Building Energy 2011, DOE 2011.

Figure 1. Continuum of residential solar system designs showing increasing integration (from left to right) with building architecture and material.

The competitiveness of BIPV in the marketplace largely depends on its cost compared with PV. We examine this issue using a bottom-up analysis of installed PV and BIPV system prices for hypothetical rooftop cases and carry this forward to estimate levelized cost of energy (LCOE) values for each case. All cost values throughout this report are provided in 2010 U.S. dollars. We also examine less-quantifiable issues that affect the development and market adoption of BIPV products.

2. BIPV CHARACTERISTICS AND GROWTH OPPORTUNITIES

As with many solar products, the market price of BIPV systems is a key factor that affects the demand for systems and resulting levels of deployment. An analysis of two California incentive programs showed that BIPV rooftop systems have been sold at higher market prices than rack-mounted PV systems. BIPV on new homes sold for about 8% more than competing PV, on average, from 2007 to 2010 (Barbose et al. 2011),[7] and the price disparity grew over the survey period, as illustrated in Figure 2. However, the prices reported in incentive program databases do not necessarily reflect downward trends in system *costs* because they are subject to a range of market dynamics.[8] Higher BIPV system prices may result from supply chain issues for products and services or consumers' willingness to pay premiums. Incentives may also influence the price disparities between rack-mounted PV and BIPV. In Section 4, we discuss the cost differentials between PV and BIPV in detail.

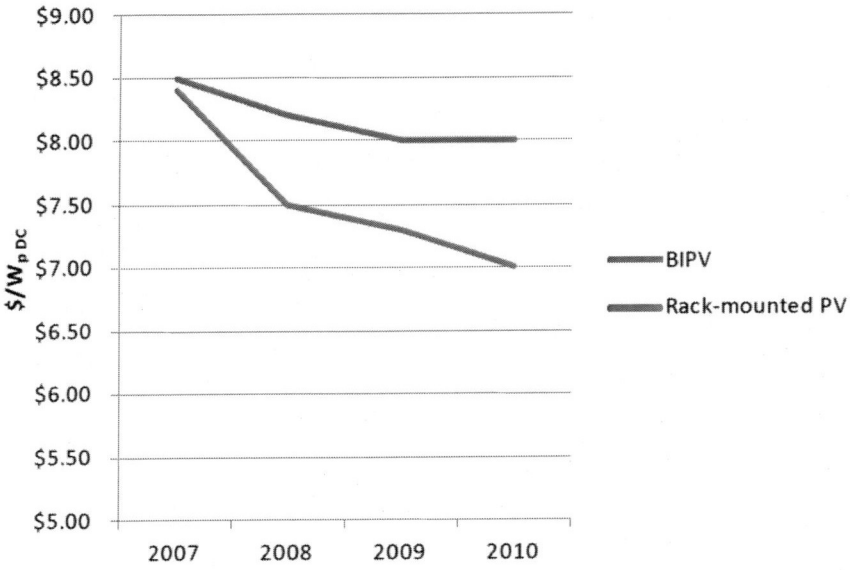

Source: Barbose et al. 2011.

Figure 2. Average installed system prices of rooftop PV and BIPV systems (2- to 3-kW systems) on newly constructed homes in the United States.[9]

BIPV may hold potential to increase PV-suitable space on buildings. One study of PV supply curves found that building rooftops in the United States could host about 660 GW of installed capacity, assuming the installation of rack-mounted PV with a 13.5% conversion efficiency (Denholm and Margolis 2008).[10] This assessment of PV-suitable rooftop areas accounted for shading, obstructions, and architectural designs that cannot accommodate traditional module form factors. Arguably, BIPV could increase these PV-suitable areas on buildings if products are lightweight or designed for specific building features. The International Energy Agency (IEA) estimated that incorporating BIPV on building façades could increase PV-suitable surfaces by about 35% (IEA 2002). Yet, there is considerable uncertainty about these findings, including how PV-suitable spaces are defined and how the lower energy generation potential of PV devices on vertical building surfaces reduces the economic viability of projects. Appendix E provides more information on these points.

BIPV's aesthetic advantages over traditional PV could increase consumer appeal and provide growth opportunities. Additional considerations about BIPV market factors, such as industry interest and government support, are listed in Table 1.

Table 1. Potential Opportunities for BIPV Market Growth

Installation cost reductions	• Lower non-module costs – elimination of racking hardware, and greater use of traditional roofing labor and installation methods • Cost offsets for displacing traditional building materials • Lower supply chain costs – leverage more established channels to market
Improved aesthetics	• Consumer willingness to pay premiums in some markets • Broader appeal for residential solar product designs
Higher technical potential	Increased PV-suitable space on buildings
Solar industry interest	• Showcase applications • High growth potential • Technology differentiation may help suppliers distinguish themselves • Possible cost reductions and new channels to market
Government support	• Maintain historic/cultural building designs • BIPV-specific incentives in select international markets

3. HISTORY AND STATUS OF BIPV DEVELOPMENT AND DEPLOYMENT

In the late 1970s, the U.S. Department of Energy (DOE) began sponsoring projects to advance distributed PV systems, including collaborations with industry to integrate PV with building materials. By the 1980s, companies such as General Electric, Solarex, and Sanyo had developed PV shingle prototypes, but technical challenges and high costs slowed the commercialization of these products (SDA and NREL 1998).[11] As PV technologies became increasingly efficient and reliable in the years that followed, more stakeholders pursued the blending of PV devices with building materials. In 1993, DOE initiated a program called Building Opportunities in the United States for PV (PV:BONUS), which was designed, in part, to help commercialize innovative BIPV products (Thomas and Pierce 2001). Similar programs were established by groups in Europe and Japan around the same time (Arthur D. Little 1995).[12] Today, partnerships among PV manufacturers, architects, and building-materials suppliers intend to address barriers and bring new cost-competitive products to the market (Fraile et al. 2008).

Because BIPV has been known mostly for showcasing solar applications in sustainable building designs, it has been regarded as a niche product compared to rack-mounted PV products. One of the first U.S. homes with BIPV was built in 1980 (Arthur D. Little 1995), and systems were later incorporated on commercial structures such as the 4 Times Square Building in New York City in 2001, where about 15-kW of amorphous silicon (a-Si) BIPV was installed (DOE 2001). Larger BIPV systems have been installed more recently, including a 6.5-$MW_{p\,DC}$ system on the Hongqiao Railway Station in China, completed prior to the 2010 Shanghai World Expo (IEA 2011). At the simplest level, BIPV systems are derivatives of common PV module designs and installation methods; early product designs were often highly customized for specific buildings and architectural features. Today, BIPV products have more standardized designs that are intended to integrate with many common building materials. Although the market prices for BIPV are still higher than for rack-mounted PV (see Section 2), new products offer lower costs and better performance than BIPV systems of the past.

Overall, the global deployment of BIPV is small in comparison with the deployment of rack-mounted PV. By some estimates, the cumulative installed capacity of BIPV (and related semi-integrated PV products) worldwide was 250–300 MW by the end of 2009 (EuPD Research 2009, Pike Research 2010). This was about 1% of the cumulative installed capacity of distributed PV systems at that time (Mints and Donnelly 2011). Part of this limited market share can be attributed to the price premium of BIPV relative to rack-mounted PV, as well as qualitative factors we discuss in the following sections.

4. RESIDENTIAL SYSTEM PRICE ANALYSIS: BIPV CASES AND 2010 PV SYSTEM BENCHMARK

Our analysis approximates the cash purchase prices (or overnight capital costs) for three hypothetical BIPV systems and a typical rack-mounted PV system (2010 benchmark price) installed on residential rooftops in the United States. We develop a granular perspective on cost factors that underlie reported system prices, which may help guide strategic decisions by research and development managers and policymakers. This bottom-up method of estimating BIPV system prices disregards the pricing parameters determined by markets, focusing instead on objective inputs as a means to assess cost-reduction opportunities and challenges.

The National Renewable Energy Laboratory (NREL), in collaboration with industry, developed the methodology we use to analyze system costs. This method is similar to the approach used by many solar project developers to approximate the book value of solar assets, characterizing the unsubsidized cash purchase price for residential systems. Our analysis includes all of the materials, labor, regulatory costs, and overhead and profit (O&P) margins for installed residential systems. Costs are provided in terms of 2010 U.S. dollars per peak watt of DC PV capacity ($/W or $/W$_{P\,DC}$).

We analyze the prices of a typical rack-mounted PV rooftop system and three BIPV rooftop systems (Table 2). The PV and BIPV cases do not represent specific commercial products. Assumptions are intended to represent typical system technologies and costs for 2010, unless otherwise noted. Efficiency assumptions are based on commercial modules and reasonable expectations about BIPV derate factors; we assume that residential BIPV systems have more inactive areas (i.e., areas that are not converting sunlight to electricity such as frames) than rack-mounted PV. Details about efficiency assumptions are provided in Appendix A.

The PV Reference Case is based on the most commonly deployed PV technology in the world, crystalline silicon (c-Si) modules. The BIPV cases include a derivative of the PV Reference Case, as well as two examples using thin-film technologies. To assess the cost implications of flexible form factors, we consider that the cells of BIPV Thin-film Case 2 are packaged with flexible materials. The other cases have rigid form factors.

Table 2. Summary of Cases Used to Analyze Residential Rooftop PV and BIPV System Prices

Scenario	Technology	Form	Efficiency	Module Area (m^2)
PV Reference Case	c-Si	Rigid	14.5%	1.28
BIPV Derivative Case	c-Si	Rigid	13.8%	0.58
BIPV Thin-film Case 1	CIGS	Rigid	11.2%	0.58
BIPV Thin-film Case 2	a-Si	Flexible	5.8%	0.58

a-Si—amorphous silicon; CIGS—Cu(In,Ga)Se$_2$; c-Si—crystalline silicon.

Module dimensions affect installation labor costs because of the time required to affix and wire systems. Smaller modules with the same form factor as larger modules generally result in higher labor costs for systems; the rate-determining step is often clamping and through-roof mounting. In our BIPV cases, we assume that the product dimensions for BIPV are smaller than those of traditional PV and more comparable to traditional roofing shingles. These dimensions allow for use of traditional roofing techniques (i.e., the BIPV installers may use nails and hammers; electricians are only needed to complete the wiring and interconnection). We assume the PV Reference Case modules are 1.28 m^2 (0.808 m × 1.580 m), typical of 2010 industry standards (Photon International 2011), and that electricians install the PV modules, in addition to completing the wiring and interconnection. It should be noted, however, that the share of installation labor from electricians varies widely across the United States, and it is not necessarily a BIPV-specific advantage to use general contractors to install modules. We also assume that all PV module surface area is exposed to the sun. We assume our BIPV shingles layer like traditional asphalt shingles so that some areas remain unexposed. We assume that only the exposed areas contain PV devices, and that this area is 0.58 m^2 (1.411 m × 0.411 m) per module. Including the layered areas, we assume the BIPV product's total area is about 0.80 m^2, which is between the sizes of traditional asphalt shingles and residential PV modules.[13] The dimensions of these BIPV shingles are also similar to the dimensions of today's BIPV products,[14] supporting the notion that they can be installed using traditional roofing methods.

For the purposes of cost modeling, solar system sizes are based on the following area constraints: 35 m^2 for the PV Reference Case and 40 m^2 for the BIPV cases. Area assumptions for the BIPV cases are slightly higher because smaller modules could potentially increase access to PV-suitable areas. According to these assumptions and the module efficiencies listed in Table 2, system capacities are about 5.0 kW for the PV Reference Case, 5.7 kW for the BIPV Derivative Case, 4.7 kW for the BIPV Thin-film Case 1, and 2.5 kW for the BIPV Thin-film Case 2. The cost analysis is normalized in terms of $/W$p_{DC}$, including an analysis of traditional building materials, to enable direct comparisons among the cases.

Residential-sector system costs vary owing to a number of factors, including channels to market, installer experience, and differences in regional labor rates, permitting fees, and taxes. A recently released technical report by NREL, which uses the same methodology as this report to estimate system book values, found a 2010 residential benchmark price of $5.71/$W_{p\ DC}$ (Goodrich et al. 2011). Accounting for the regional factors that affect system costs, this 2010 benchmark price has a standard deviation of about 8%.

Product designs and intended functionality create inherent cost differences between PV and BIPV. BIPV devices often include additional materials such as flashing to ensure buildings are protected from a wide range of weather conditions. On the other hand, most BIPV products reduce installation costs by eliminating common PV mounting hardware such as struts, z-channels, and clips and associated labor costs. BIPV modules may also install more quickly than incumbent PV modules. Additionally, it is important to consider the potential cost benefits of offsetting the use of traditional building materials (e.g., asphalt shingles) in the areas where BIPV is installed.

4.2.1. Installation Costs

The installation cost differences between the rack-mounted PV benchmark case and BIPV cases are driven by the installation methods and materials requirements of each system. The most significant installation cost difference for the BIPV cases is from the elimination of racking hardware and associated labor costs. Racking components serve as the direct interface between PV modules and rooftop structures, and our BIPV cases eliminate the need for this robust material interface.

Secondly, we assume that BIPV's functionality as a roofing material, along with its smaller module size, would allow project developers to use lower-cost roofing contractors to install the products using traditional roofing methods.[15] In this regard, the BIPV cases have cost advantages because we assume that electricians are only needed for the final wiring steps and commissioning; we also assume that BIPV shingles would take 65% less time to install on a per-module basis. Because the BIPV modules are smaller, however, more total time is required to install the modules on a per-area basis.

We include overhead and profit (O&P) margins and sales taxes as part of the installation cost category, and these factors are determined as a percentage of total system costs. We assume that all of the cases have the same rates for O&P and taxes (54% overhead, 30% profit, and 5% sales tax). Yet, because system costs differ for each case, the resulting O&P and sales tax costs are not the same. We also assume that "indirect capital costs" such as permitting fees, which vary considerably across jurisdictions, are equivalent for all of the cases: $900 per system, based on interviews with U.S. industry stakeholders. The resulting proportion of indirect capital costs to system costs is a function of system capacity, which differs for each case. Therefore, larger systems have smaller indirect capital costs in terms of $/W.

We approximate 2010 installation cost differences between comparable PV and BIPV residential systems. Quantifying prospective cost-reduction opportunities is beyond the scope of this analysis; however, we recognized that new installation methods and system designs may likely affect the cost differentials illustrated in our analysis. Cheaper mounting structures and faster installation methods for rack-mounted PV, for example, would reduce the installation cost advantages of the BIPV cases. It is also possible that novel and integrated circuitry could lower BIPV wiring costs in the future, yet it is uncertain whether the benefits would be specific to BIPV or diffuse across the sector, reducing PV system costs as well. As such, we limit our analysis to the costs of currently available technologies assuming that much of the bill-of-materials is the same across all cases. These materials include inverters, meters, system monitors, AC/DC disconnects, combiner boxes, fuses and holders, conduit, and wiring.

4.2.2. Module Costs

We assume that modules in the BIPV cases have the performance of roofing products in addition to the functionality of PV devices, and we assume that this is achieved by the use of more materials than are used in incumbent PV modules. We recognize that materials are highly specific to BIPV product designs, which may include novel framing, flashing, adhesives, and materials to mitigate heat gains. Different BIPV module designs can also lead to trade-offs between module costs and installation costs.

Because of the variability of BIPV module designs, we simplified the assumptions of the BIPV cases by adding 10% premiums to the costs of commercially available PV modules, adding about $0.15–$0.22/W depending on the technology. The BIPV module mark-up accounts for all materials (framing, flashing, etc.) that are necessary to enable safe rooftop installations, and it includes base layers installed directly onto roof decks and under BIPV modules. Felt paper barriers or wraps, for instance, are typically installed under asphalt shingles, although some BIPV products require more expensive materials made of polypropylene or elastomeric sheets. When purchased in volume, the installed prices of the higher-end materials are about $2.00–3.00/m^2 (RSMeans 2010), adding a cost of about $0.02/W, which we assume is part of the 10% premium. Given that the BIPV module premium is a rough estimate, we include a range of BIPV module costs in the uncertainty analysis, which is available in Appendix C.

4.2.3. Flexible Packaging Costs

There is on-going debate about the value of flexible PV and BIPV products, but it is clear that BIPV does not require flexible form factors. Among currently installed BIPV and semi-integrated PV designs, the most widely deployed products are rigid and use c-Si technologies (Ceron et al. 2010).[16] There may be specific opportunities for flexible BIPV products such as on commercial buildings, where roofing materials are often flexible, or for niche applications such as military tents or buildings with tensile fiberglass roofs (e.g., the Denver International Airport). Opportunities may also exist in the residential rooftop sector for non-planar products such as Stiles. In all cases, the potential benefits of flexible form factors must be weighed against added costs and performance considerations. The thin-film technologies analyzed in this report (CIGS and a-Si) can be developed into flexible BIPV products, and there are some advantages: flexible modules tend to have lower weight than glass modules (up to 90% lighter), and they can have lower shipping and installation costs. Another advantage is that they may better accommodate building areas with limited structural support.[17] These potential advantages, however, may not compensate for the additional costs of flexible barrier materials or additional challenges involving long-term product safety, reliability, and durability. Different tolerances to long-term ultraviolet radiation exposure, for instance, can affect anti-soiling properties, transmissivity, and the adhesive stability of materials, which can significantly affect device performance (Kempe 2009). We only examine the costs of flexible cell packaging in the a-Si BIPV case (Thin Film Case 2). We assume that a-Si requires top sheets with water vapor transmission rates (WVTRs) of about 10^{-2} g/m^2/day, which are available for about $10/m^2. This would add about $0.40/W compared to standard glass-glass packaging. BIPV Thin-film

Case 1 (CIGS) is modeled as a rigid product, although CIGS can be developed into flexible form factors. CIGS is highly sensitive to water, and it requires higher performing barriers than a-Si technologies with WVTRs of less than 10^{-4} g/m^2/day (Leffew et al. 2011). The prices of these barriers are expected to decrease substantially with industry scale-up; however, they are costly today at $40–$80/m^2. Assuming a $40/m^2 top sheet barrier film and $20/m^2 back sheet, we estimate that flexible CIGS modules are about $0.60/W more expensive than glass-glass CIGS modules. Our analysis only models CIGS BIPV as a rigid product, so this packaging premium does not appear in the results.

4.2.4. Building Material Cost Offsets

If BIPV products completely replace traditional building materials, overall system costs should reflect a commensurate cost offset. Developing multifunctional products is a central challenge for BIPV product designers because building materials often require higher durability than PV devices, and BIPV must meet codes and standards for both PV and building products. 18 We assume these challenges have been overcome for the BIPV cases in this paper. The costs and performance of standard roofing materials vary. Asphalt shingles are the most common product installed on U.S. residential rooftops; they account for more than 50% of U.S. residential sector market share (National Roofing Contractors Association 2011 b). For most conditions, asphalt shingles last about 17–20 years, and installed costs are between $18–$32/m^2 (RSMeans 2010). More expensive rooftop products such as clay tiles may last more than 50 years and often provide better insulation and fire protection than less costly products. Figure 3 illustrates the range of retail prices for fully installed products in 2008–2009. To estimate the value of potential offsets for BIPV, we converted roofing product prices from $/m^2 to $/W, accounting for module dimensions and efficiencies. Table 3 lists the values for several roofing materials to illustrate general cost trends. Later, we estimate asphalt shingle offset values according to the technology characteristics of each BIPV case (Section 4.3.2).

PV products have a range of efficiencies, and lower-efficiency products require more space than higher-efficiency products for equivalent system power capacities. Similarly, lower-efficiency BIPV technologies require more space and displace more traditional products than higher-efficiency BIPV technologies; thus, in terms of $/W, offsets are inversely related to PV efficiencies: a 6.3%-efficient device has more than double the offset value of a 14.5%-efficient device for an equivalent roofing product. Table 4 lists the approximate offset values for selected technologies and building materials, illustrating the possible range of residential offset values by highlighting a low-case offset (shingles) and a high-case offset (clay tiles).

Table 3. Average Installed Retail Prices for Traditional Residential Roofing Materials, Converted to $/W Based on the BIPV Derivative Case (13.8%-efficient, 0.58 m^2)

Roofing Product	$/m^2	$/W
Asphalt shingle	$25.08	$0.18
Wood shingle	$51.13	$0.37
Concrete tile	$57.86	$0.42
Slate tile	$78.58	$0.57
Metal tile	$101.45	$0.74
Clay tile	$116.52	$0.85

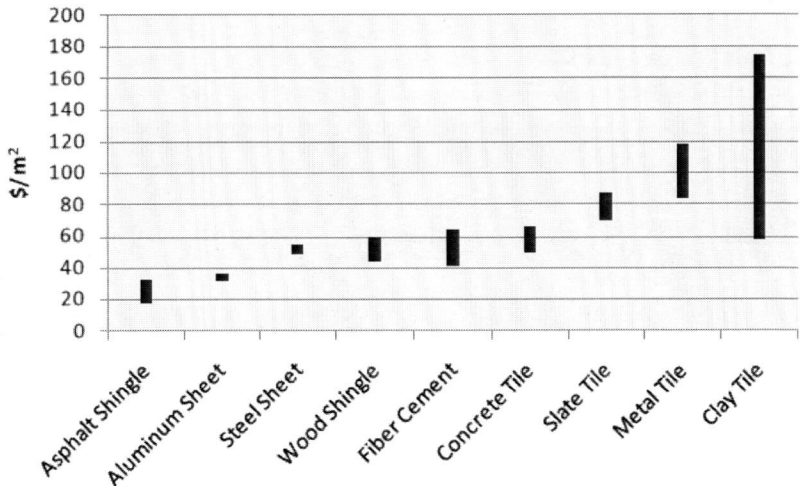

Source: RSMeans 2010.

Figure 3. Installed retail prices of residential roofing products in the United States, 2008–2009.

Table 4. Estimated Offset Values for the Residential BIPV Cases

Technology	PV metrics		Residential Material Offsets ($/W)	
	Efficiency	W_p/m^2	Asphalt Shingle	Clay Tile
a-Si	5.8%	58	$0.43	$2.01
CIGS	11.2%	113	$0.22	$1.03
c-Si	13.8%	138	$0.18	$0.85

The following sections summarize the installed system price estimates for the BIPV cases compared with the PV Reference Case. Installed system prices account for all component costs, installation labor costs, indirect capital costs, sales taxes, and margins. Operations and maintenance costs are not included. Offsets for BIPV cases are denoted as "Offset shingles." The listed "Effective price" values include the relative offsets for asphalt shingles in the BIPV cases.

4.3.1. Detailed Results for PV Reference Case vs. BIPV Derivative Case

Cost gaps between the PV Reference Case and BIPV Derivative Case are mostly from differences in module costs and installation costs, as well as the value of offsets from traditional building materials. These cost differentials also impact channel costs, as illustrated in Figure 7.

Our analysis shows that the effective price of the BIPV Derivative Case is $0.69/W lower than the price of the PV Reference Case, a difference of more than 10%. An offset of $0.18/W was included because we assume the BIPV case replaces asphalt shingles. Elimination of racking hardware and associated labor is estimated to reduce total BIPV costs by $0.55/W ($0.27/W for labor and $0.27/W for materials). Not all differences reduce the BIPV case's costs, however. The smaller module dimensions for the BIPV case result in $0.08/W higher total module-related installation labor costs (despite non-electrician labor rates), and the 10% BIPV module mark-up adds $0.22/W. However, even with these increases, net installation costs are less for the BIPV Derivative Case. These lower costs help to reduce the costs of O&P and sales taxes. Indirect capital costs are also slightly lower in the BIPV Derivative Case because its system capacity (5.7 kW) is larger than the PV Reference Case (5.0 kW).[20] Figure 5 shows all of the major cost categories for the two c-Si cases.

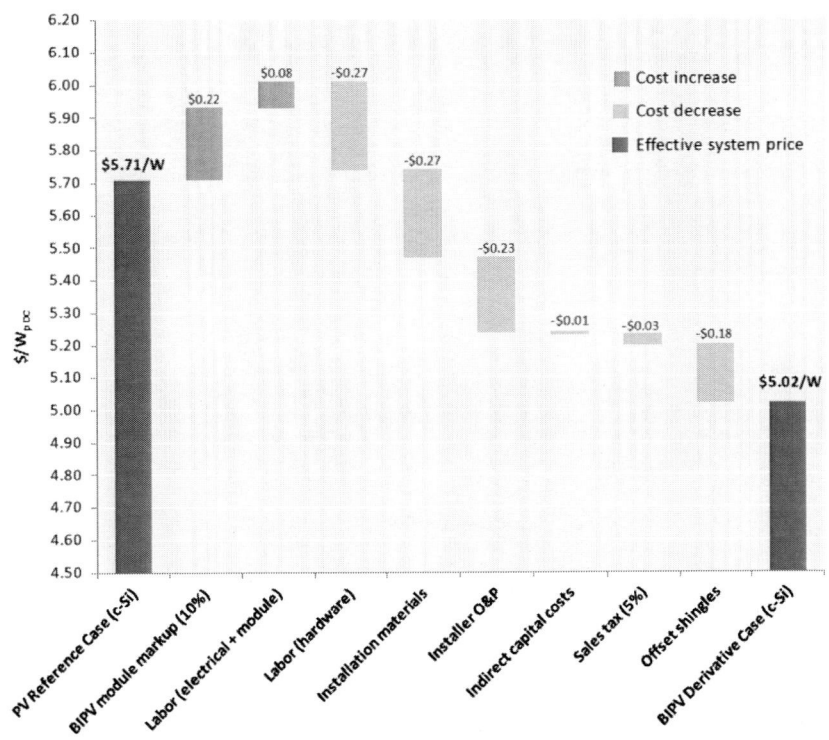

Figure 4. Price differences between the PV Reference Case (c-Si, 14.5% efficiency, 2010 benchmark price) and the BIPV Derivative Case (c-Si, 13.8% efficiency).[19]

The context of the illustrated BIPV price advantages is critical to the understanding of general opportunities and challenges for BIPV products. In today's market, few BIPV products are fully integrated with building materials as described in the BIPV cases of this report; therefore, the hypothetical BIPV cases are essentially *near-term possibilities* that are compared with the 2010 benchmark PV system price. Because PV system prices continue to decrease, soon-to-be commercialized BIPV products are chasing a moving target. In addition, this report's analysis assumes that the BIPV cases benefit from the cost advantages of manufacturing products on a similar scale as rack-mounted PV products. As discussed previously, the costs associated with converting a PV device into a BIPV product remain highly uncertain. We chose a 10% module mark-up for the BIPV cases, but this mark-up could be much higher, adding to overall channel costs such as installer O&P. For these reasons, BIPV cases in this report are not representative of today's BIPV system prices.

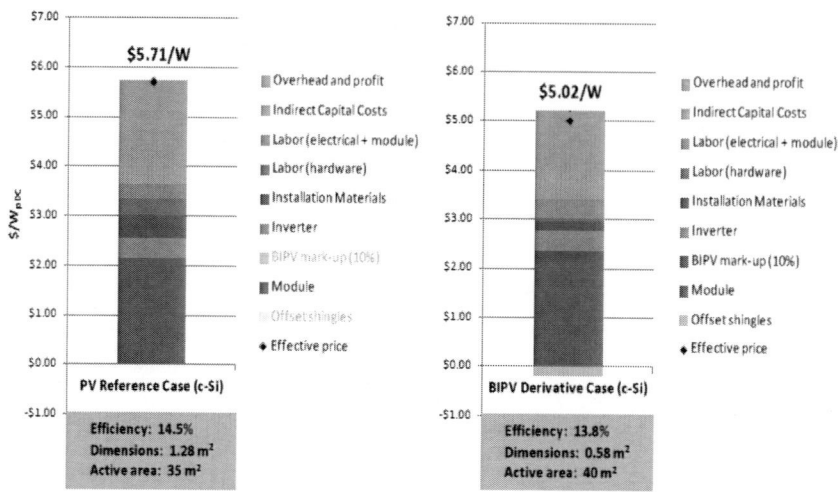

Note: The listed BIPV price includes shingle cost offsets, shown as a negative bar in the figure on the right.[21]

Figure 5. Comparison of residential rooftop prices for the PV Reference Case and the BIPV Derivative Case.

4.3.2. Summary Results for All Cases

Figure 6 illustrates the system price differences between the PV Reference Case and the BIPV cases. The effective prices of the BIPV Derivative Case and BIPV Thin-film Case 1 are about the same, with prices that are more than 10% lower than the PV Reference Case. The effective price of BIPV Thin-film Case 2 is about 1% less expensive than the PV Reference Case. BIPV Thin-film Case 2 is the only flexible product shown in Figure 6, and potential benefits of flexible form factors are not accounted for in the figure.

Among the three BIPV cases, installed system prices vary as a result of module costs and efficiencies. Module costs are representative of module spot prices in December 2010 with a 10% distributor margin (Photon International 2011). The magnitudes of the other mark-ups, as well as offset values, are also affected by module costs and system efficiencies. Appendix C provides more information about the estimated system prices for the BIPV cases in this report, using a Monte Carlo analysis to address the uncertainties about a number of assumptions.

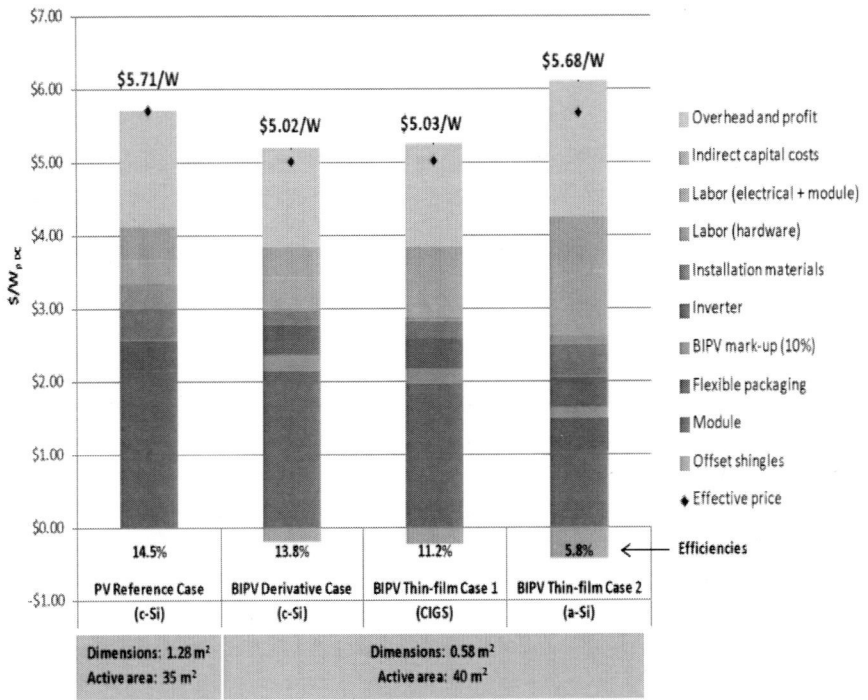

Note: Listed BIPV prices include building-material cost offsets (shown as negative bars in the figure).

Figure 6. Comparison of residential rooftop prices for the PV Reference Case and three BIPV cases.

5. SYSTEM PERFORMANCE AND LEVELIZED COST OF ENERGY

The sections above summarize residential rooftop PV and BIPV system costs in terms of price per unit of installed capacity ($/W). To understand the economic viability of solar systems, costs must also be understood in terms of the levelized cost of energy (LCOE), which is based on a system's installed price, its total lifetime cost, and its lifetime electricity production. The following sections address LCOE.

It is unlikely that BIPV systems will perform as well as rack-mounted PV products, and, although models have been developed using empirical data from select products, performance changes for BIPV are not easy to generalize

(Neises 2011). In some cases, BIPV has shown higher average operating temperatures, thermally accelerated degradation (e.g., corrosion of metallization), and increased soiling on low-sloped roofs (Schams and TamizhMani 2011). Novel products require stringent testing under a range of environmental conditions, yet lacking lifetime performance data can complicate the process of developing appropriate product warranties. For an analysis of LCOE, we only consider the relative losses that result from higher average cell temperatures for BIPV compared with off-roof mounted PV.[22]

The operating temperatures of rooftop solar systems are affected by several parameters, including installation configurations, ambient temperatures, irradiance, and wind speeds. As such, assessing BIPV system lifetime performance is complex and highly dependent on technologies, installation designs, and local environmental conditions (Neises 2011). Air gaps under modules improve system performance because convective currents help cool modules, and studies have suggested mounting PV arrays 3–6 inches from roof decks in order to optimize power output, cooling, and wind loading (Dunlop 2007). However, by the nature of BIPV designs, most integrated systems are mounted directly onto building surfaces with no air gaps. Compared to off-roof mounted systems, products installed directly onto building roof surfaces have shown performance losses as high as 7% in high-temperature environments such as Arizona, where module temperatures can reach 95°C (Schams and TamizhMani 2011).[23] In cooler climates, performance losses from heating are less of an issue: one study of two PV-integrated tile systems in Colorado found that a product mounted on counter battens (about 1 inch of air space) produced 3.4% more power than an identical tile system mounted directly on the roof deck (Muller et al. 2009).

To account for the relative performance losses of BIPV systems, which we consider to be mounted directly onto building surfaces with minimal underside airspaces, we examine LCOE using the installed price assumptions listed in Figure 6. To estimate LCOE ranges, we analyze two U.S. locations with different solar resource conditions (Tucson and Boston) and consider that all systems are south-facing with a 25-degree tilt and a derate factor of 85%. Financing costs and tax benefits are included in the calculation, and we use a typical structure of 80% financing of system prices with a 30-year mortgage and weighted average capital costs (WACC) of 5.9%. We assume a nominal discount rate of 10.8%. Other assumptions are summarized in Table 5.

Table 5. Selected Performance Assumptions for the PV and BIPV Cases

Scenario	Technology	Rated Efficiency	Temperature Coefficient (Pmpp(%/°C)[24])	Annual Degradation
PV Reference	c-Si	14.5%	-0.49	0.5%
BIPV Derivative	c-Si	13.8%	-0.49	0.5%
BIPV Thin-film 1	CIGS	11.2%	-0.45	1.5%
BIPV Thin-film 2	a-Si	5.8%	-0.21	1.0%

Estimates of the unsubsidized LCOE values for the cases in this report are illustrated in Figure 7. These values were derived using NREL's System Advisor Model (SAM),[25] which includes options to estimate performance losses for BIPV based on module mounting structure inputs and weather data.[26] For most locations and PV technologies, SAM estimates a performance loss of 2.0%–4.5% (relative) between open-rack systems and close-mount systems; this assumes that differences in module edges and framing designs do not affect heat transfers (Neises 2011).

The spread of LCOE estimates for the four scenarios shown in Figure 7 varies from the range of installed system prices owing to differences in module efficiencies, degradation rates, and performance in moderate (Boston) and high-temperature (Tucson) environments.[27] The BIPV LCOE values range from 7% lower to 5% higher than the PV Reference Case. The greatest BIPV installed price advantages (about 12% for the BIPV Derivative Case and BIPV Thin-film Case 1) are reduced by performance disadvantages. The most temperature-tolerant BIPV case, BIPV Thin-film Case 2 (a-Si), maintains its cost disadvantages, although it is more competitive in a high-temperature environment like Tucson. The LCOE analysis highlights how higher degradation rates and performance disadvantages can affect the economic viability of some technologies and system designs. Commensurate with our assumptions, this analysis shows that LCOE values of c-Si BIPV and CIGS BIPV could be competitive with c-Si rack-mounted PV in most environments. For the purposes of additional discussion, a sensitivity analysis of LCOE and system prices for the BIPV Derivative Case (c-Si) is provided in Appendix D.

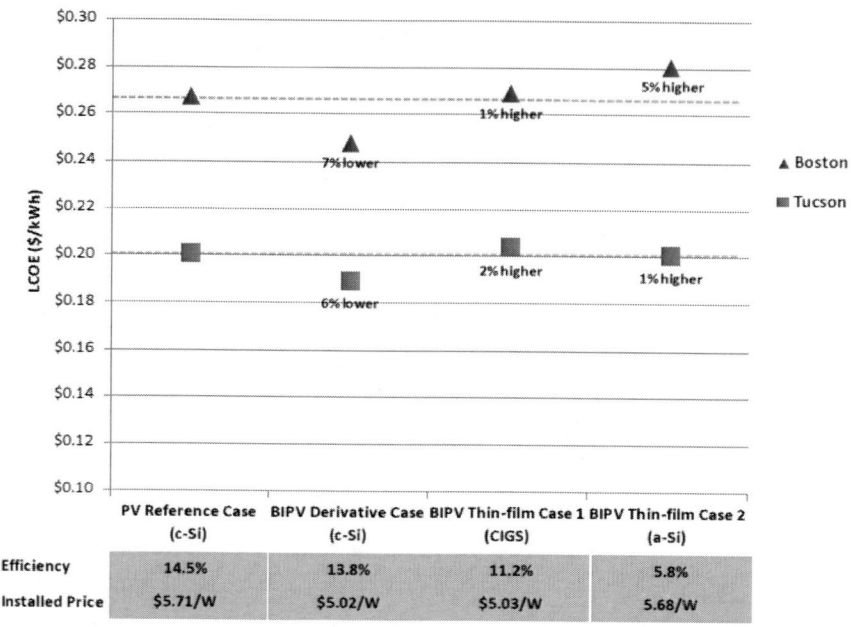

Note: Listed percentages illustrate LCOE differences relative to the PV Reference Case.[28] The LCOE calculations are based on consistent system price and financing structure assumptions for both locations—Boston and Tucson—but account for differences in estimated system prices, efficiencies, temperature coefficients, and degradation rates. All systems are south-facing and tilted at 25 degrees.

Figure 7. Unsubsidized U.S. residential rooftop LCOE values for the BIPV shingle cases compared with the PV Reference Case.

6. Qualitative Considerations for BIPV

The limited deployment of BIPV worldwide is likely a result of higher system prices (see Section 2), but other factors may also affect market opportunities. Especially when cost differences between PV and BIPV are modest, less-quantifiable issues may impact growth rates for the BIPV sector. Factors germane to BIPV include the following:

- **The value of aesthetics**

  One of BIPV's core advantages is that it might better address the aesthetic interests of many stakeholders. Market data have shown some level of consumer willingness to pay premiums for BIPV (Barbose et al. 2011), yet it is too early to determine how consumers value aesthetics and whether this tolerance of premium prices will recede or expand as solar markets evolve. Most importantly, it is unclear to what degree aesthetics will be a driving force for widespread deployment of PV technologies.

  Although it is not possible to assess objectively whether a PV system is attractively designed, previous experiences suggest that aesthetics matter to PV consumers. In Oradell, New Jersey, for example, PV panels recently installed on municipal electric poles have drawn public criticism for looking "ugly" (Navarro 2011). BIPV responds to these types of concerns with designs that blend or are otherwise visually consistent with traditional building materials (Zanetti 2010). To assess a number of design criteria, one recent survey of more than 170 building professionals resulted in guidelines about the quality of PV integration into building structures (Probst and Roecker 2007); BIPV case studies have also provided insights about specific design opportunities (Eiffert and Kiss 2000). Incorporating architectural considerations into solar product designs might help marginally higher-priced BIPV systems vie for market share, although BIPV's perceived aesthetic advantages might be challenged by PV mounting structures that are increasingly designed to appeal to similar architectural interests. Overall, the behavioral economic drivers of PV adoption are not well understood, but there may be substantial opportunities for aesthetically designed BIPV systems if costs are competitive with other technologies.

- **Codes and standards**

  The landscape of codes and standards issues is more complex for BIPV than for competing PV products. PV devices and roofing products have different durability, safety, and performance requirements, and certification processes for PV and BIPV are often disconnected (Sharma and Herron 2011). Standards bodies are working to harmonize codes and expand BIPV-specific guidelines. However, navigating codes and standards issues may continue to be more complex for BIPV.

- **Policy and regulatory issues**

There have been increasing policy opportunities to support BIPV through tailored incentive programs, measures to promote sustainable buildings, and efforts to address building codes and other relevant regulations. Targeted, price-driven incentive programs such as premium-rate feed-in tariffs for BIPV can increase market opportunities, as assessments of experiences in Italy and France have suggested (EuPD Research 2009). Solar access laws can also create environments that are amenable to BIPV products. Switzerland has building regulations, for instance, that serve to protect the cultural nature of architectural designs, and these regulations inhibit PV installations unless they are integrated into building envelopes (Zanetti 2010). In this regard, historic preservation commissions and code officials can influence the designs of residential PV systems. Solar access laws and local regulatory policies vary (DSIRE 2011), and some regulatory structures present barriers for all types of residential solar applications (Starrs et al. 1999). BIPV might help overcome some of these regulatory barriers, and growth opportunities are likely to remain strongly linked to policy schemes.

- **Market segmentation**
The dynamics of BIPV adoption might differ from those of rack-mounted PV because products are often designed for more discrete market opportunities such as new residential roofs or specific building products (e.g., clay tiles). PV modules are not necessarily fungible across sectors (residential, commercial, and utility), but they are much more transferrable than BIPV products, where designs vary greatly even within sectors. Although BIPV opportunities are more segmented, they must compete with the costs of a much more robust rack-mounted PV industry. Limiting installations to smaller markets can affect the ability to achieve cost targets through manufacturing scale-up.

CONCLUSION

Although the deployment of BIPV is relatively low, opportunities remain promising. Decreasing module costs, increasing consumer interest in solar energy, and policy schemes that support distributed generation systems have the potential to increase rates of BIPV market growth. The commercialization of solar products that have the full functionality of building materials has been very limited, but systems are increasingly being developed to account for design aesthetics and installation-cost reductions. This continuum of integration is leading to more solar products that may fully replace traditional building materials. Significant challenges have affected product development and market adoption of BIPV over the past 30 years, and several barriers remain. Despite high interest from solar energy stakeholders, substantial research and development efforts, and policy support in some markets, BIPV and semi-integrated PV products accounted for less than 1% (250–300 MW) of global installed capacity of distributed systems in 2009. A primary reason for BIPV's limited deployment is that the average market price of installed systems is currently higher than for rack-mounted PV. However, our findings support the notion that BIPV could have competitive and, in some cases, lower installed system prices than rack-mounted PV, by reducing installation costs and offsetting traditional building materials. Lower-efficiency PV products have higher offset values than higher-efficiency products, although these gains may be reduced by higher overall system prices. In the best case, effective system prices for c-Si BIPV and CIGS BIPV could be more than 10% lower than for rack-mounted c-Si PV. However, across all the cases analyzed in this report, the opportunities for BIPV to reduce solar system prices may become more limited as the prices of rack-mounted PV continue to decrease. If installed system prices are lower for BIPV, then select products may also have marginally lower LCOE values than PV depending on anticipated performance losses from the higher average operating temperatures that result from systems mounted directly onto rooftop surfaces. Accounting for higher installed system prices and performance issues, the LCOEs for a-Si thin-film BIPV likely will exceed those for rack-mounted c-Si PV in the residential rooftop sector. The value of aesthetic designs and flexible module form factors is not well understood, but these system characteristics may help mitigate the disadvantages of higher prices for select products. BIPV faces more complex product development issues and market adoption dynamics than rack-mounted PV, and these issues may significantly impede progress to capitalize on price-reduction opportunities. The key issues that may limit market growth include new system performance considerations, more discrete market opportunities, and greater testing requirements to ensure relevant codes, standards, and warranty issues are addressed appropriately. BIPV-specific incentives are not widespread, but they have increased deployment of BIPV systems in some regions. In the near-term, the more complex technology and design issues and relatively small-scale production capacity of BIPV likely may result in continued price disadvantages compared with rack-mounted PV systems. In this regard, the success of many residential rooftop BIPV products may hinge on the aesthetic value of product designs and a consumer willingness to pay premiums for non-traditional systems. Our analysis supports the notion that BIPV has the potential to reduce the installed system prices of comparable rack-mounted PV in residential rooftop markets. Market experiences suggest, however, that realizing these opportunities can be challenging.

REFERENCES

Arthur D. Little, Inc. (1995). Building-Integrated Photovoltaics (BIPV): Analysis and U.S. Market Potential. Prepared for the U.S. Department of Energy's Office of Building Technologies.

Barbose, G.; Darghouth, N.; Wiser, R.; Seel, J. (2011). Tracking the Sun IV: An Historical Summary of the Installed Cost of Photovoltaics in the United States from 1998 to 2010. Berkeley, CA: Lawrence Berkeley National Laboratory.

Building Energy. (2011). Building Energy website. http://www.building energyvt.com/assets/Uploads/Solar-PV/schumann-montpelier-solar-pv.JPG, accessed 2011.

CPUC (California Public Utilities Commission). (2010). California Solar Initiative Program Handbook. www.gosolarcalifornia.ca.gov/documents/ CSI_HANDBOOK.PDF, accessed May 2011.

Ceron, I.; Olivieri, L.; Caamano-Martin, E.; Neila, J. (2010). "State of the art of BIPV technologies." Paper presented at Solar Building Skins Energy Forum, Bressanone, Italy.

Denholm, P.; Margolis, R. (2008). Supply Curves for Rooftop PV-Generated Electricity for the United States. NREL/TR-6A0-44073. Golden, CO: National Renewable Energy Laboratory.

DOE (U.S. Department of Energy). (2001). "4 Times Square New York City – Highlighting Performance." DOE Office of Building Technology. State and Community Programs. http://apps1.eere.energy.gov/buildings/ publications/ pdfs/commercial_initiative/29940.pdf, accessed March 2011.

DOE. (2011). Solar Multimedia website. www.eeremultimedia.energy accessed 2011.

DSIRE (Database of State Incentives for Renewables and Efficiency). (2011). Incentives/Policies for Solar: Rules, Regulations & Policies. www.dsire usa.org/solar/incentives/index.cfm?EE=1&RE=1&SPV=1&ST=1&searcht ype=Acces s&solarportal=1&sh=1, accessed 2011.

Dunlop, J. (2007). "National Joint Apprenticeship and Training Committee (NJATC) for the Electrical Industry." In Photovoltaic Systems. Orland Park, IL: American Technical Publishers.

Easton, P.; Wild, J.; Halsey, R.; McNally, M. (2010). Financial Accounting for MBAs – 4th Edition. Cambridge Business Publishers.

Eiffert, P.; Kiss, G. (2000). Building-Integrated Designs for Commercial and Institutional Structures – A Sourcebook for Architects. U.S. Department of Energy's Office of Power Technologies, Photovoltaics Division, and the Federal Energy Management Program. www1.eere.energy accessed January 2011.

EuPD Research. (2009). "Range of BIPV-Applications Rise." www.eupd-research.com, accessed March 2011.

EPIA (European Photovoltaic Industry Association). (2010). SOLARIS Newsletter. www.epia.org/publications/solaris-newsletter/editorial.html, accessed April 2011.

Fraile, D.; Despotou, E.; Latour, M.; Slusaz, T.; Weiss, I.; Caneva S.; Helm, P.; Goodal, J.; Fintikakis, N.; Schellekens, E. (2008). "PV Diffusion in the Building Sector." European Photovoltaic Industry Association, Sunrise Project.

Goodrich, A.; James, T.; Woodhouse, M. (2011). Residential, Commercial, and Utility-Scale Photovoltaic (PV) System Prices in the United States: Current Drivers and Cost-Reduction Opportunities. In preparation. Golden, CO: National Renewable Energy Laboratory.

IEA (International Energy Agency). (2002). Potential for Building Integrated Photovoltaics. Technical Report. PVPS T7–4.

IEA. (2011). Clean Energy Progress Report. www.iea.org/papers/2011/CEM_Progress_Report.pdf, accessed April 2011.

Kempe, M.D. (2009). "Ultraviolet light test and evaluation methods for encapsulants of photovoltaic modules." Solar Energy Materials & Solar Cells. Elsevier B.V. ISSN: 0927-0248. www.elsevier.com/locate/solmat, accessed May 2011.

Leffew, K.W.; Boussaad, S.; Reardon, D.; Tate, H.; Glassmaker, N.; Dean, D.; Nunes, G.; Zhao, C.Q.; Samuels, S.L.; Schneider, J.; Carcia, P.; Mclean, R. (2011). "Systems Approach to High Performance CIGS Material Set Including Flex Ultra-Moisture Barrier and Hi-Temp MLI Substrate." DuPont Research and Development presentation at the DOE/NREL Photovoltaic Module Reliability Workshop 2011. www1.eere.energy accessed May 2011.

Luma Resources. (2010). Luma Resources Solar Shingle Datasheet. www.lumaresources.com/datasheets/lrss.pdf, accessed March 2011.

Mints, P.; Donnelly, J. (2011). Analysis of Worldwide Markets for Solar Products & Five-Year Application Forecast 2010/2011. Navigant Consulting, Inc.

Muller, M.T.; Rodriguez, J.; Marion, B. (2009). "Performance Comparison of a BIPV Roofing Tile System in Two Mounting Configurations." NREL/CP-520-45948. Golden, CO: National Renewable Energy Laboratory.

National Roofing Contractors Association. (2011 a). "Asphalt Shingles." www.nrca.net/consumer/types/asphalt.aspx, accessed May 2011.

National Roofing Contractors Association. (2011 b). Interview. April 2011.

Navarro, M. (2011). "Solar Panels Rise Pole by Pole, Followed by Gasps of 'Eyesore'." New York Times. www.nytimes.com/2011/04/28/science/earth/28solar.html?_r=3&hp, accessed April 2011.

Neises, T. (2011). Development and Validation of a Model to Predict the Temperature of a Photovoltaic Cell. Master of Science Thesis. University of Wisconsin–Madison.

Paidipati, J.; Frantzis, L.; Sawyer, H.; Kurrasch, A. (2008). Rooftop Photovoltaic Market Penetration Scenarios. NREL/SR-581-42306. Golden, CO: National Renewable Energy Laboratory.

Photon International. (2011). "All about modules." Solar Power Magazine, February 2011.

Pike Research. (2010). Research Report: Building-Integrated Photovoltaics. Boulder, CO: Pike Research.

Probst, M.M.; Roecker, C. (2007). "Towards an improved architectural quality of building integrated solar thermal systems (BIST)." Solar Energy 81 (2007) 1104–1116.

PV News. (2010). Vol. 30, No. 4, April 2010. Boston, MA: Greentech Media.

RSMeans. (2010). Building Construction Cost Data. Norwell, MA: Reed Construction Data.

Schams, B.; TamizhMani, G. (2011). "BAPV Modules with Different Air Gaps: Effect of Temperature on Relative Energy Yield and Lifetime." Presented at IEEE's 37th Photovoltaic Specialist Conference (PVC).

SDA and NREL (Solar Design Associates and the National Renewable Energy Laboratory). (1998). Photovoltaics in the Built Environment: A Design Handbook for Architects and Engineers. Prepared on behalf of the U.S. Department of Energy.

Sharma, S.; Herron, D. (2011). "Reliability and Application Challenges for Flexible Thin-Film (BIPV) Modules." United Solar Ovonic presentation at PV Module Reliability Workshop, Golden, CO, February 17, 2011.

SEIA and GTM Research (Solar Energy Industries Association and Greentech Media Research). (2010). U.S. Solar Market Insight – 2010 Year-In-Review. SEIA and GTM Research.

Starrs, T.; Nelson, L.; Zalcman, F. (1999). Bringing Solar Energy to the Planned Community – A Handbook on Rooftop Solar Systems and Private Land Use Restrictions. Prepared for the U.S. Department of Energy. Contract Number: DE – FG01 – 99EE10704.

SunPower. (2010). SunPower SunTile Datasheet. SPR-76RE-BLK-U.

Thomas, H.P.; Pierce, L.K. (2001). "Building Integrated PV and PV/Hybrid Products – The PV:BONUS Experience." NREL/CP-520-31138. Golden, CO: National Renewable Energy Laboratory.

White House. (2011). "The State of the Union 2011: Winning the Future." www.whitehouse.gov/state.

Wilcox, S.; Marion, W. (2008). Users Manual for TMY3 Data Sets. NREL/TP-581-43156. Golden, CO: National Renewable Energy Laboratory.

Zanetti, I. (2010). "Sustainable renovation of historical buildings concepts for solar integration." Paper presented at the Solar Building Skins Energy Forum, Bressanone, Italy.

APPENDIX A: BACKGROUND ON EFFICIENCY ASSUMPTIONS

This report uses 2010 commercial modules as reference points for assumptions about the BIPV cases. Because we assume that the BIPV shingles are smaller than incumbent PV modules and likely include additional materials and extra spacing, we assume slightly lower efficiencies. These assumptions are listed in Table 6. Examples of typical 2010 commercial module efficiencies are listed in Table 7.

Table 6. Summary of Efficiency Assumptions for the PV and BIPV Cases

Case	Module Dimensions (m^2)	Cell efficiency	Module Derate	Module efficiency
PV Reference Case (c-Si)	1.28	16.7%	87.0%	14.5%
BIPV Derivative Case (c-Si)	0.58	16.7%	83.0%	13.8%
Thin-film BIPV Case 1 (CIGS)	0.58	14.0%	80.0%	11.2%
Thin-film BIPV Case 1 (a-Si)	0.58	7.3%	80.0%	5.8%

Table 7. Examples of 2010 Commercial Modules for c-Si, CIGS, and a-Si PV Technologies[29]

PV Technology	Product	Module Dimensions (m²)	Cell-to-Module Derate	Module Efficiency
c-Si	Residential-sector modules			
	Suntech: STP185S - 24/Adb+	1.28	87.3%	14.5%
	SunPower: E19/320	1.63	-	19.6%
CIGS	Residential-sector modules			
	MiaSolé: MS140GG	1.07	-	13.1%
	MiaSolé: MS120GG	1.07	-	11.2%
	Q-Cells: QSMART 75	0.76	-	9.9%
	Q-Cells: QSMART 95	0.76	-	12.5%
	Solar Frontier: SF140-L	1.23	-	11.4%
	Solar Frontier: SF155-L	1.23	-	12.6%
a-Si	Residential- and commercial-sector modules (flexible)			
	Uni-Solar: PVL-68	1.12	83.4%	6.1%
	Uni-Solar: PVL-136	2.16	86.6%	6.3%
	Xunlight: XR-12 1117 A	1.64	-	5.9%
	Xunlight: XR-36 1117 A	4.72	-	6.2%

APPENDIX B: COST INPUT TABLES

Estimated system prices in this report are generated from various inputs, including component costs (e.g., modules, racking hardware, inverters), labor rates, channel costs (i.e., margins), and indirect costs (e.g., commissioning fees, taxes). Inputs vary according to supply and demand, regional issues, project scale, and a number of other factors. In this sense, costs reflect a snapshot of market dynamics for a given period. The costs input assumptions listed below (Table 8 through Table 13) represent U.S. averages for residential rooftop systems in 2010.

Table 9. Assumptions of Indirect Costs

Permitting & Commissioning	$900/system
Sales Tax	5%

Table 8. Assumptions of Mark-ups

Module distributor mark-up	10%
Materials mark-up	30%
Inverter mark-up	15%
Installer overhead	54%
Installer profit mark-up	30%

Table 10. Material Costs and Installation Labor Requirements for the PV Reference Case

Material Category	Component costs ($/W)	Installation labor allocation requirements			
		Units	Units/system	Electrical (hours/unit)	General (hours/unit)
Module	2.15*	Modules	27	0.2	
Inverter	$0.42	Inverters	1	4	2
Wiring	$0.03	Linear feet (ft)	237†	0.05	
Other electrical‡	$0.19	Electrical subsystem	1	4.5	
Mounting hardware	$0.37	Module racks	27		1.4
Total materials cost	$3.16				
Total installation labor requirements				25.8	39.8

* Ex-factory gate price ($1.95/W, 2010 Photon).
+ retail margin (10%) = $2.15/W.
† Total wiring (237 ft) = home run wiring (77 ft)
+ row to combiner wiring (160 ft)
‡ "Other electrical" includes: meter, system monitor, and disconnects.

Table 11. Material Costs and Installation Labor Requirements for the BIPV Derivative Case

Material Category	Component Costs ($/W)	Installation labor allocation requirements			
		Units	Units/system	Electrical (hours/unit)	General (hours/unit)
Module	2.37*	Modules	68		0.07
Inverter	$0.42	Inverters	1	4	2
Wiring	$0.07	Linear feet (ft)	541†	0.05	
Other electrical‡	$0.17	Electrical subsystem	1	4.5	
Mounting hardware	$0.00	Module racks	0		0
Total materials cost	$3.03				
Total installation labor requirements				35.6	6.8

* Ex-factory gate price ($1.95/W, 2010 Photon) + retail margin (10%) + BIPV mark-up (10%) = $2.37/W † Total wiring (541 ft) = home run wiring (141 ft) + row to combiner wiring (400 ft) ‡ "Other electrical" includes: meter, system monitor, and disconnects.

Table 12. Material Costs and Installation Labor Requirements for the BIPV Thin-film Case 1

Material Category	Component costs ($/W)	Installation labor allocation requirements			
		Units	Units/system	Electrical (hours/unit)	General (hours/unit)
Module	2.17*	Modules	68		0.07
Inverter	$0.42	Inverters	1	4	2
Wiring	$0.09	Linear feet (ft)	541†	0.05	
Other electrical‡	$0.21	Electrical subsystem	1	4.5	
Mounting	$0.00	Module racks	0		0

	Component costs	Installation labor allocation requirements			
hardware					
Total materials cost	$2.89				
Total installation labor requirements				35.6	6.8

* Ex-factory gate price ($1.79/W, 2010 Photon) + retail margin (10%) + BIPV mark-up (10%) = $2.17/W † Total wiring (541 ft) = home run wiring (141 ft) + row to combiner wiring (400 ft) ‡ "Other electrical" includes: meter, system monitor, and disconnects.

Table 13. Material Costs and Installation Labor Requirements for the BIPV Thin-film Case 2

	Component costs	Installation labor allocation requirements			
Material Category	($/W)	Units	Units/system	Electrical (hours/unit)	General (hours/unit)
Module	1.65*	Modules	68		0.07
Inverter	$0.42	Inverters	1	4	2
Wiring	$0.17	Linear feet (ft)	541†	0.05	
Other electrical‡	$0.41	Electrical subsystem	1	4.5	
Mounting hardware	$0.00	Module racks	0		0
Total materials cost	$1.00				
Total installation labor requirements				35.6	6.8

* Ex-factory gate price ($1.36/W, 2010 Photon) + retail margin (10%) + BIPV mark-up (10%) = $1.65/W † Total wiring (541 ft) = home run wiring (141 ft) + row to combiner wiring (400 ft) ‡ "Other electrical" includes: meter, system monitor, and disconnects.

APPENDIX C: INSTALLED SYSTEM PRICE UNCERTAINTY ANALYSIS

The analysis of PV and BIPV systems in this report relies on a number of assumptions, including national average labor rates. We recognize that installed residential system prices vary across the United States, and that there are significant uncertainties in our assumptions, such as the module costs for the BIPV cases. Labor costs, component costs, site-specific costs (e.g., permitting and taxes) and supply chain costs (operating O&P margins) differ across regions. Incentives and the scale and experience of companies can impact these factors; thus, it is difficult to compare the costs of specific projects or to generalize the costs of systems without including margins of error. This Monte Carlo analysis provides insights into the factors that most contribute to uncertainties of the BIPV price analysis results in this report. Information about the uncertainties of the PV Reference Case (2010 PV system benchmark price) is available in the NREL report by Goodrich et al. (2011). The following uncertainty analysis, summarized in Figure 8 through Figure 10 and Table 14 through Table 16, is based on factors that are most likely to vary across projects. Values listed are considered reasonable for 2010 based on published data and installer-reported information. The most frequently reported information is listed as "mode." Triangular distributions were assumed for all variables. Because the BIPV cases are hypothetical and less defined by the market, we include a relatively broad assessment of module efficiencies, module prices, and module sizes that can impact installed system prices. Offsets for traditional building materials are excluded. The factors affecting system prices that we noted earlier, such as module efficiencies and system sizes, are particularly relevant to this uncertainty analysis (see Section 4.2 *Major Cost Differential Categories*) along with the listed range of input assumptions (Table 14 through Table 16).

Table 14. Assumptions for the Monte Carlo Simulation of the BIPV Derivative Case

Module		min	mode	max
[1] Module efficiency	@STC, 1000 W/m^2	13.0%	13.8%	19.0%
[2] Module price	per W$_{PDC}$	$2.00	$2.37	$2.60
[3] Module size	m^2	0.50	0.58	1.28
Installation Labor				
[4] Electrical	$ per hour	$16.66	$49.00	$81.34
[4] General construction	$ per hour	$11.25	$33.10	$54.95
[5] Labor content (all types)	hours	48.5	64.7	80.9
[6] Operating overhead		25%	54%	65%
[7] Profit on labor		10%	30%	35%
Inverter				
[†] Inverter price	per W$_{PDC}$	$0.25	$0.42	$0.65
Installation Materials				
[†] Mounting hardware	per module	$52.29	$69.71	$89.25
[†] Wiring, conduit, connectors	per module	$3.60	$4.80	$6.14
[†] Supply chain costs	%-materials price	15%	30%	35%
Site work				
[†] Permitting		$0.00	$0.00	$500.00
[†] Grid Interconnect		$0.00	$900.00	$2,000.00

[1] Non-exhaustive survey of standard c-Si module datasheets, Sunpower E18 / 400 datasheet

[2] Beate Knoll, "Downward path", Module Price Survey, Photon International, January 2011; Jeremy Heron, "Shining the Light", Photon International, September 2010 (20% gross margin assumption)

[3] Non-exhaustive survey of standard c-Si module datasheets, Sunpower E18 / 400 datasheet

[4] U.S. BLS, National average labor rate (electrical contractor), min/max, 2009

[5] Private conversations with U.S. installers (labor hours by component; ±25% productivity variation based on installer experience, site specifics)

[6] Average operating overhead (16%), electrical contractor (annual billings >$4MM), Electrical Contractor Handbook, RS Means, 2010

[7] Average profit (10%), electrical contractor (annual billings >$4MM), Electrical Contractor Handbook, RS Means, 2010

[†] 2010-2011 NREL (authors) private conversations with installers (review of confidential project cost data provided by installers under Non-Disclosure Agreements)

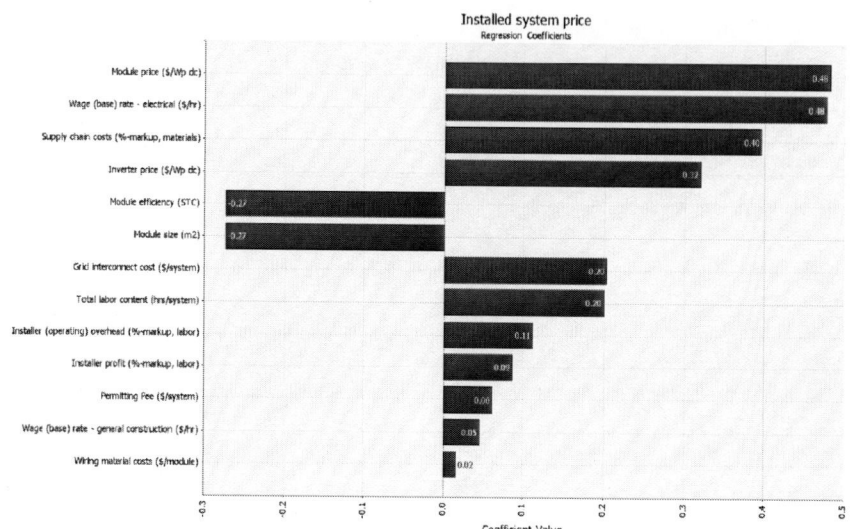

Figure 8. Residential system price Monte Carlo analysis results, probability distribution function, for the BIPV Derivative Case.

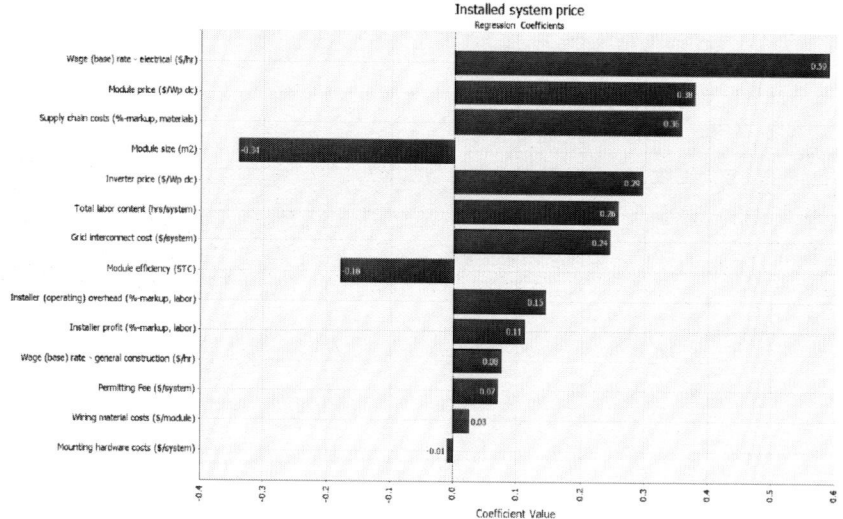

Figure 9. Residential system price Monte Carlo analysis results, probability distribution function, for the BIPV Thin-film Case 1.

Table 15. Assumptions for the Monte Carlo Simulation of the BIPV Thin-film Case 1

Module		min	mode	max
[1] Module efficiency	@STC, 1000 W/m^2	10.0%	11.2%	12.5%
[2] Module price	per W_{PDC}	$1.90	$2.17	$2.40
[3] Module size	m^2	0.50	0.58	1.28
Installation Labor				
[4] Electrical	$ per hour	$16.66	$49.00	$81.34
[4] General construction	$ per hour	$11.25	$33.10	$54.95
[5] Labor content (all types)	hours	48.5	64.7	80.9
[6] Operating overhead		25%	54%	65%
[7] Profit on labor		10%	30%	35%
Inverter				
[†] Inverter price	per W_{PDC}	$0.25	$0.42	$0.65
Installation Materials				
[†] Mounting hardware	per module	$52.29	$69.71	$89.25
[†] Wiring, conduit, connectors	per module	$3.60	$4.80	$6.14
[†] Supply chain costs	%-materials price	15%	30%	35%
Site work				
[†] Permitting		$0.00	$0.00	$500.00
[†] Grid Interconnect		$0.00	$900.00	$2,000.00

[1] Non-exhaustive survey of standard c-Si module datasheets, Sunpower E18 / 400 datasheet

[2] Beate Knoll, "Downward path", Module Price Survey, Photon International, January 2011; Jeremy Heron, "Shining the Light", Photon International, September 2010 (20% gross margin assumption)

[3] Non-exhaustive survey of standard c-Si module datasheets, Sunpower E18 / 400 datasheet

[4] U.S. BLS, National average labor rate (electrical contractor), min/max, 2009

[5] Private conversations with U.S. installers (labor hours by component, ±25% productivity variation based on installer experience, site specifics)

[6] *Average operating overhead (16%), electrical contractor (annual billings > $4MM)*, Electrical Contractor Handbook, RS Means, 2010

[7] *Average profit (10%), electrical contractor (annual billings > $4MM)*, Electrical Contractor Handbook, RS Means, 2010

[†] 2010-2011 NREL (authors) private conversations with installers (review of confidential project cost data provided by installers under Non-Disclosure Agreements)

Table 16. Assumptions for the Monte Carlo Simulation of the BIPV Thin-film Case 2

Module		min	mode	max
[1] Module efficiency	@STC, 1000 W/m^2	5.0%	5.8%	6.5%
[2] Module price	per W$_{PDC}$	$1.40	$1.65	$1.90
[3] Module size	m^2	0.50	0.58	1.28
Installation Labor				
[4] Electrical	$ per hour	$16.66	$49.00	$81.34
[4] General construction	$ per hour	$11.25	$33.10	$54.95
[5] Labor content (all types)	hours	48.5	64.7	80.9
[6] Operating overhead		25%	54%	65%
[7] Profit on labor		10%	30%	35%
Inverter				
[†] Inverter price	per W$_{PDC}$	$0.25	$0.42	$0.65
Installation Materials				
[†] Mounting hardware	per module	$52.29	$69.71	$89.25
[†] Wiring, conduit, connectors	per module	$3.60	$4.80	$6.14
[†] Supply chain costs	%-materials price	15%	30%	35%
Site work				
[†] Permitting		$0.00	$0.00	$500.00
[†] Grid Interconnect		$0.00	$900.00	$2,000.00

[1] Non-exhaustive survey of standard c-Si module datasheets, Sunpower E18 / 400 datasheet

[2] Beate Knoll, "Downward path", Module Price Survey, Photon International, January 2011;
   Jeremy Heron, "Shining the Light", Photon International, September 2010 (20% gross margin assumption)

[3] Non-exhaustive survey of standard c-Si module datasheets, Sunpower E18 / 400 datasheet

[4] U.S. BLS, National average labor rate (electrical contractor), min/max, 2009

[5] Private conversations with U.S. installers (labor hours by component, ±25% productivity variation based on installer experience, site specifics)

[6] *Average operating overhead (16%), electrical contractor (annual billings >$4MM)*, Electrical Contractor Handbook, RS Means, 2010

[7] *Average profit (10%), electrical contractor (annual billings >$4MM)*, Electrical Contractor Handbook, RS Means, 2010

[†] 2010-2011 NREL (authors) private conversations with installers (review of confidential project cost data provided by installers under Non-Disclosure Agreements)

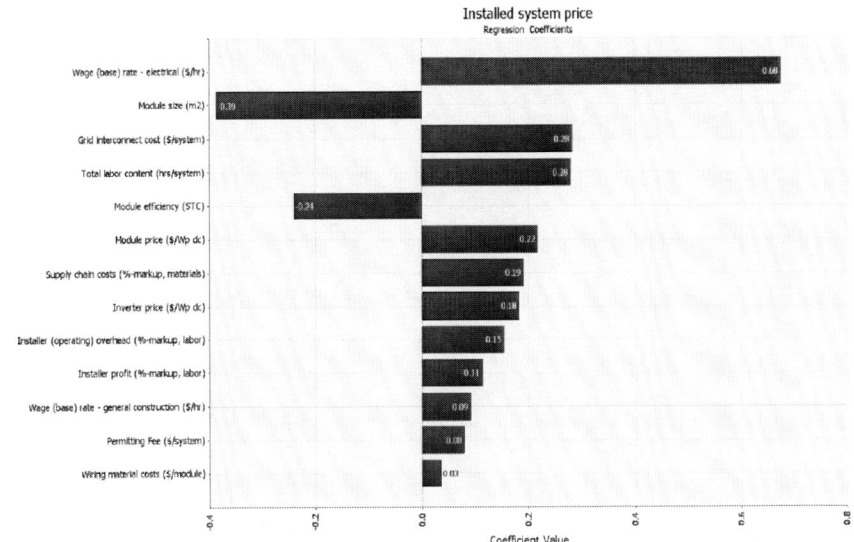

Figure 10. Residential system price Monte Carlo analysis results, probability distribution function, for the BIPV Thin-film Case 2.

APPENDIX D: LEVELIZED COST OF ENERGY AND SYSTEM COST PARAMETRIC

LCOE estimates are sensitive to many variables including solar resources, site-specific weather conditions, capital costs, and operations and maintenance costs. The LCOE values in this report were calculated using SAM (version 2011.5.23), which is available for download on the NREL website.[30] Performance losses for BIPV cases were estimated using the options for select mounting structures in SAM's "Simple Efficiency Module."[31]

Because this report focuses on an assessment of installed system prices, which contain uncertainties, a parametric analysis of LCOE values and installed system prices for the BIPV Derivative Case is provided below (Figure 11).

The simulation is for a system located in Tucson, and it is financed by a mortgage loan with a 5.9% WACC without incentives and a nominal discount rate of 10.8%.

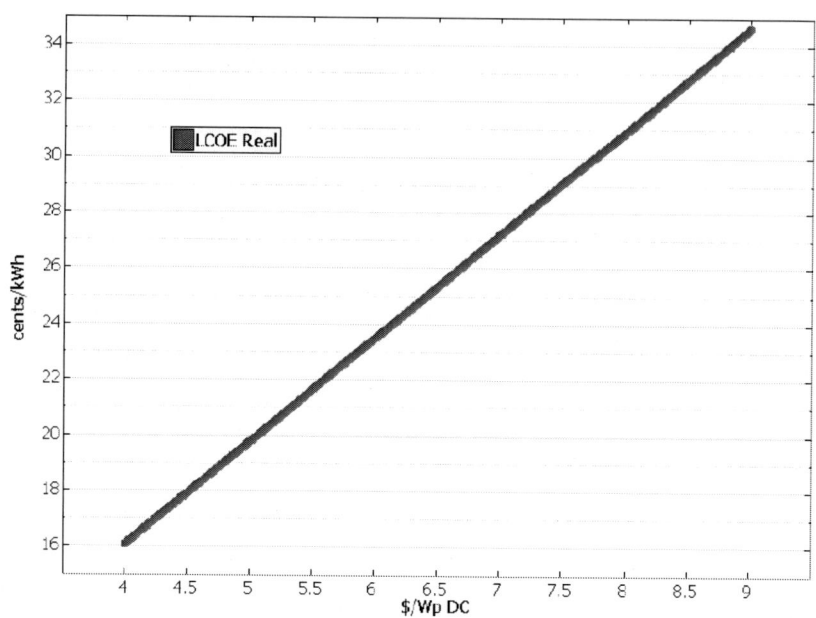

Figure 11. Sensitivity analysis of LCOE and installed system prices for the BIPV Derivative Case.

APPENDIX E: TECHNICAL POTENTIAL OF BIPV ON U.S. BUILDING SURFACES

A number of studies assess how building surfaces could be used to generate electricity from PV devices. Most studies focus on small areas or cities, and, owing to a range of assumptions, national estimates cannot be extrapolated easily. Our analysis of the technical potential for PV on buildings is mostly based on one national study by NREL in combination with a report by IEA.

A 2008 NREL study quantified supply curves for PV-generated electricity under three scenarios. Using building data from 2007, the study estimated that rooftops could host about 660 GW of PV capacity (350 GW residential, 310 GW commercial) from 13.5%-efficient PV modules (Denholm and Margolis 2008). Total roof space estimates were developed using building data from McGraw-Hill and the U.S. Energy Information Administration's 2005 Residential Energy Consumption Survey (RECS) and 2003 Commercial Building Energy Consumption Survey (CBECS).

Estimating PV-suitable spaces on building surfaces is a key factor in determining the technical potential of PV. The NREL study cited a 2008 report from Navigant Consulting, Inc. (NCI) about PV-penetration scenarios. Because suitable areas vary by region, average national estimates are uncertain and subject to a range of assumptions. Climate conditions and roof designs, obstructions from HVAC equipment, shading from adjacent structures and vegetation, and weight limitations can all affect suitability (Paidipati et al. 2008). NCI's study divided the United States into two climate zones, considering the southernmost states—including California, Nevada, Georgia, South Carolina, and Hawaii—to be warm climates. The study assumed that PV-suitable space on residential rooftops is 22% of total roof areas on homes in cold climates and 27% of roof areas in warm/arid climates. For commercial buildings, NCI estimated PV-suitable space as 65% of total roof area in cold climates and 60% in warm/arid climates.[32] We assume that the differences between these numbers are due to reduced tree shading and larger HVAC units in warm/arid climates for residential and commercial buildings, respectively.

The IEA published a BIPV study in 2002 that analyzed both rooftop and façade areas for several IEA countries. The report defines PV suitability as areas that result in at least 80% of the maximum annual solar input for given slopes. In addition to shading, IEA accounted for factors such as architectural designs that would limit spaces in their assessment of six tilt angles among five types of building structures. Using this 80% solar yield criterion, IEA estimated that PV-suitable space on rooftops is about 64% (average) of total roof areas, which is similar to NCI's commercial building rooftop estimate. The IEA report estimated that suitable façade space is less than 50% of available façade areas. Applying these factors to U.S. building data, IEA estimated that PV-suitable space is about 10,000 km^2 on rooftops and about 3,800 km^2 on façades (IEA 2002). This rooftop estimate is about double the more recent NREL rooftop estimate, and it is reasonable to assume a wide margin of error in these types of national studies.

Module efficiency assumptions are critical to estimating the maximum installed power capacity for given areas. In 2010, the average efficiency of the most widely installed PV technology, c-Si, was about 14.5%. Due to the assumptions of higher derate factors for BIPV (i.e., more framing and spacing), as outlined in Appendix A, we estimated a c-Si BIPV module efficiency of 13.8%. If we use NREL's PV-suitable rooftop space estimate (4,900 km^2) and assume 13.8%-efficient BIPV modules are installed (instead of the study's 13.5% efficiency assumption), the NREL capacity estimate of 660 GW would be scaled-up to about 675 GW. If PV-suitable façade space is added, the total technical potential for BIPV could be more than 900 GW—assuming that façades would increase the estimates of PV-suitable rooftop space by about 35%, as described by IEA.

However, BIPV façade systems are likely to have lower capacity factors (CFs) than BIPV rooftop systems because of suboptimal tilt angles. Using typical meteorological year data across more than 1,000 sites in the continental United States (Wilcox and Marion 2008), CFs for south-facing c-Si rooftop systems tilted at 25 degrees range from 10% to more than 21%. A south-facing BIPV system tilted at 90 degrees (like many building façade areas) will, on average, yield about 35% less energy than an identical system tilted at 25 degrees.[33] Figure 12 and Figure 13 illustrate where CFs are most affected across the United States. The locations in northern latitudes experience smaller CF differences between 25 degrees and 90 degrees, and some southern locations maintain higher CFs at 90 degrees than systems at 25 degrees in northern areas. These CF considerations are critically important when assessing the economic viability of BIPV technologies and system designs. BIPV façade systems tilted at 90 degrees can produce electricity, but these systems will not be as economically competitive as BIPV rooftop systems in most cases in the United States.

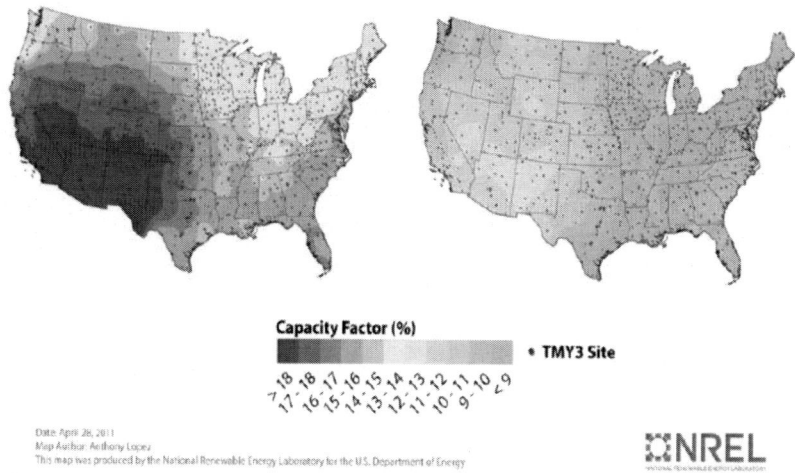

Figure 12. Regional variation of PV capacity factors for south-facing, fixed-mount systems tilted at 25 degrees (left) and 90 degrees (right).

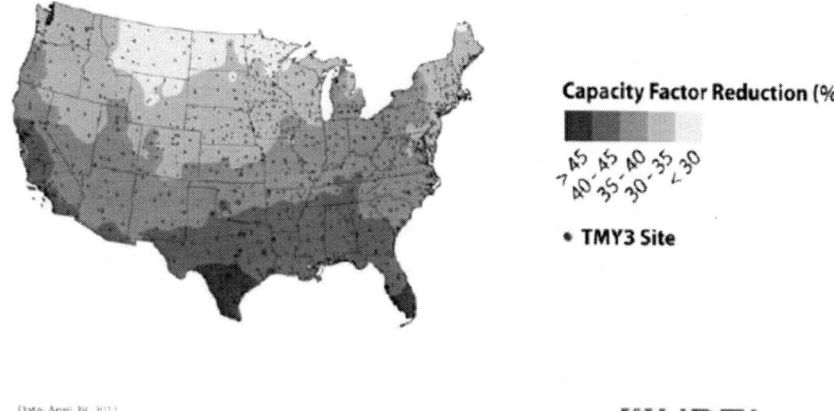

Figure 13. Relative reductions in PV capacity factors between south-facing, fixed-mount systems tilted at 25 degrees and 90 degrees.

[1] LCOE estimates do not include any federal, state, local, or utility incentives. They assume host ownership and that no taxes are paid on electricity. Mortgage payments are tax deductible.

[2] Luma Resources' solar shingle product, honored in the 2011 State of the Union Address, is composed of a polycrystalline PV module adhered to a metal shingle.

[3] Feed-in tariff (FIT) rates for small BIPV rooftop products in Italy exceed rates for rack-mounted PV by more than 10%. BIPV-specific FITs also have lower digression rates than PV technologies in some markets.

[4] BAPV is also referred to as building-adapted, -added, and -adhered PV (NanoMarkets 2010, Greentech Media 2010, Pike Research 2010, ASDReports 2010, EuPD Research 2009).

[5] Similarly, only 2% of modules eligible for California's solar-electric incentive programs are described as BIPV, and these include some partially integrated products (July 2011). The California Energy Commission's list of nearly 6,000 modules is available at: www.gosolarcalifornia.org/equipment/pv_modules.php

[6] This definition is aligned with products that qualify for France's highest BIPV "full integration" incentive, as well as a description of BIPV by the California Public Utilities Commission (CPUC 2010).

[7] These data were derived from information about systems funded through the California Solar Initiative (CSI) and New Solar Homes Partnership programs. System types (PV and BIPV) were determined using the CSI List of Eligible Modules, which lists some semi-integrated PV products as BIPV.

[8] We consider "market price" to be synonymous with "fair market value" – the value that an asset could be sold for (or an obligation discharged) in an orderly market, between willing buyers and sellers; often, but not always, it is the current market value (Easton 2010).

[9] These findings are from an analysis of about 3,000 residential rooftop PV and BIPV systems. The illustrated BIPV prices do not include any cost offsets for traditional roofing materials.

[10] Of the 660 GW, this study estimated that about 350 GW could be installed on residential rooftops and 310 GW on commercial building rooftops.

[11] BIPV system prototypes developed in 1979 and the early 1980s were evaluated at DOE-sponsored experiment stations in Massachusetts, Florida, and New Mexico.

[12] IEA PVPS Task 7. EPIA Sunrise Project (Europe). NEDO BIPV RD&D Program (Japan).

[13] Dimensions of the most common residential rooftop product (i.e., asphalt "strip shingles") are 0.29 m2 (National Roofing Contractors Association 2011 b).

[14] Luma Resources' c-Si solar shingle is 0.55 m^2 (Luma Resources 2010). SunPower's Suntile is 0.65–0.69 m2 (SunPower 2010).

[15] We used electrical contractor wage rates of $101.29/hr for the PV Reference Case and roofing contractor wage rates of $68.42/hr for all BIPV cases (RSMeans 2010).

[16] This is based on a study that assessed 238 integrated PV products produced by 109 different companies.

[17] A survey of more than 200 BAPV/BIPV products found that the majority, many of which were rigid products, weigh less than 20 kg/m^2. About 10% of the products weighed more than 30 kg/m^2, and these heavy systems may require that rooftops have additional structural support (Ceron et al. 2010).

[18] PV and rooftop building material requirements may include: IEC TC82 (WG3), FM global, ICC-ES, AC-07, FM 4470, IBC, ASCE, BOCA, SBCCI, and SFBC codes (Sharma and Herron 2011).

[19] Sources: Goodrich et al. 2011, Photon International 2011, SEIA and GTM Research 2010.

[20] In the BIPV thin-film cases, indirect capital costs are higher than in the PV Reference Case owing to smaller system capacities.

[21] Indirect capital costs include sales tax in this chart.

[22] We assume that the BIPV cases do not have higher degradation rates than comparable rack-mounted PV systems. Inputs for module degradation rates (Table 5) are specific to each technology.

[23] This study also found evidence that a system with no air space could have a 28% shorter system lifetime than a system with a 4-inch air space in high-temperature environments.

[24] Pmpp – power maximum power point. These values are from NREL's SAM (www.nrel.gov/analysis/sam/).

[25] LCOE numbers are given in terms of "real" (as opposed to nominal) dollars, and they do not include any federal, state, or local incentives.

[26] The option of "close roof mount" in SAM represents a product with minimal air flow under modules. Performance-loss estimates are within the range of performance losses observed for direct-mount BIPV products.

[27] Assessments of product quality and project risks that impact lending rates (i.e., bankability) also affect LCOEs. Novel BIPV products may have less certain performance than some PV technologies, and, as a result, financing costs could be higher. In this report, we use the same financing assumptions for all PV and BIPV cases.

[28] LCOE estimates do not include any federal, state, local, or utility incentives. They assume host ownership and that no taxes are paid on electricity. Mortgage payments are tax deductible.

[29] This information is from datasheets published by solar module manufacturers in 2011: Suntech (www.suntechpower.com), SunPower (http://us.sunpowercorp.com), MiaSolé (www.miasole.com), Q-Cells (www.q-cells.com), Solar Frontier (www.solar-frontier.com), Uni-Solar (www.uni-solar.com), and Xunlight (www.xunlight.com).

[30] www.nrel.gov/analysis/sam.

[31] SAM's "CEC Performance Model" also has the functionality to analyze BIPV systems with additional specificity about mounting configurations.

[32] A DOE BIPV report prepared by Arthur D. Little, Inc. (1995) estimated that 30% of roof areas on existing residential and commercial building rooftops would be suitable for PV.

[33] NREL's System Advisor Model.

In: Solar Photovoltaics
Editors: G. Marks and C. J. Unger

ISBN: 978-1-62257-651-7
© 2013 Nova Science Publishers, Inc.

Chapter 3

INSURING SOLAR PHOTOVOLTAICS: CHALLENGES AND POSSIBLE SOLUTIONS[*]

National Renewable Energy Laboratory

EXECUTIVE SUMMARY

Although the market for insurance products that cover photovoltaic (PV) systems is evolving rapidly, PV developers in the United States are concerned about the cost and availability of insurance. Annual insurance premiums can be a significant cost component, and can affect the price of power and competition in the market. Moreover, the market for certain types of insurance products is thin or non-existent, and insurers' knowledge about PV systems and the PV industry is uneven. PV project developers, insurance brokers, underwriters, and other parties interviewed for this research identified specific problems with the current insurance market for PV systems in the United States and suggested government actions that could facilitate the development of this market through better testing, data collection, and communication.

Insurance premiums make up approximately 25% of a PV system's annual operating expense. Annual insurance premiums typically range from 0.25% to 0.5% of the total installed cost of a project depending on the geographic location of the installation. PV developers report that insurance costs comprise

---

[*] This is an edited, reformatted and augmented version of National Renewable Energy Laboratory Technical Report, No NREL/TP-6A2-46932, dated February 2010.

5% to 10% of the total cost of energy from their installations, a significant sum for a capital-intensive technology with no moving parts.

Because insurance is purchased annually and future premiums are uncertain, developers who offer fixed-price contracts for the electric output of their systems must use estimated future insurance costs. Developers generally pass the risk of higher future insurance premiums on to their customers through higher escalation rates or other contract elements, thereby increasing the cost of solar electricity to entities that host PV systems on their property under power purchase agreements.

The fledgling nature of the renewable energy industry makes obtaining affordable insurance challenging. These challenges include insurers' unfamiliarity with PV technologies, a lack of historical loss data (i.e., insurance claims), and limited test data for the long-term viability of PV products under real-life conditions. The lack of information and insight about the solar PV industry contributes to perceived risk associated with the technology and installation techniques among insurance underwriters and brokers, which leads to higher premiums than would likely prevail in a more mature market. Finally, the PV industry's ongoing innovation in contractual structures and business models necessitates corresponding innovation in insurance products to match the industry's requirements.

Our research identified several areas for action on the part of the federal government, national laboratories, and other stakeholders to support and accelerate the development of insurance products for PV technologies and systems:

- **Expand Availability of PV Historical Loss Data**
  Centrally assembled and evaluated data for historical insurance claims for PV systems would allow the insurance industry to better gauge the risks associated with PV installations. A large database—potentially comprising several proprietary industry data sets—that includes parameters related to system operation, availability, and insurance loss would be highly useful for the insurance industry in assessing risk and setting competitive rates. The U.S. Department of Energy (DOE) or the National Renewable Energy Laboratory (NREL) could represent an objective third-party data aggregator.
- **Evaluate Expansion of Renewable Energy Business Classification**
  Insurance professionals interviewed for this report indicated current business classification systems are insufficient to gauge various risks of renewable energy businesses. The Standard Industrial

Classification (SIC) and North American Industry Classification System (NAICS) codes combine most renewable energy industries into a narrow range of classifications. Additional delineation—perhaps through an incremental coding system—would allow insurance underwriters to better assess insurance claims related to workers' compensation, operating loss, and other risks highly specific to the renewable energy industry.

- **Develop Module and Component Testing Capabilities and Services Offered by Federal Labs**
  The PV industry would benefit from better access to detailed testing procedures for assessing PV systems for vulnerability to weather-related stresses, including severe wind, hail, and extreme temperatures. NREL should coordinate with other national laboratories and testing facilities to ensure that advanced testing capabilities are available to PV module and inverter manufacturers and system integrators.

- **Advance Industry Standards for Installers**
  At present, there are guidelines but no standards to ensure the competency of PV system installers. The DOE has been involved in developing these guidelines. However, the insurance industry interviewed for this analysis clearly communicated that unified state or federal standards could reduce risk associated with PV installation and operation and could thus lead to lower insurance premiums for solar developers and the commercial, industrial, and government entities that host PV installations on their rooftops or land. Extended reliance on the North American Board of Certified Energy Practitioners (NABCEP) certification process could improve quality and reduce accidents, and could thus lead to reduced insurance premiums. In addition, opening communication among the insurance industry, the DOE, and the development community could lead to better understanding of PV systems, improved guidelines, and lower costs.

1. INTRODUCTION

Allocation of risk is the principal concern of the insurance industry. Individuals, families, and corporations buy insurance to protect against financial damages or loss of property or persons. Insurance underwriters provide insurance to protect against potential damage or loss based on the associated risks.

According to the Insurance Information Institute, risk is: "[t]he chance of loss to the person or entity that is insured" (Ins. Info. Inst. 2009a). In each situation, the actual risk depends upon what is being insured, what events might occur, and how likely each occurrence might be. For example, geography and weather often play key roles in contributing to potential risks. Therefore, the insurance industry assesses the total risk they are covering for any particular situation and determines the likelihood that they will have to make a payment if that risk becomes an actual loss.

The annual insurance premium payment is determined based on the potential for the risks to become actual losses. For insurance companies to be profitable, the premiums they receive for insuring against losses must be greater than the actual payouts in the event of losses occurring.

Investments in PV are often viewed by underwriters as quite risky for two main reasons: the technologies are newer (i.e., most systems do not have a long history of operational data) and there are fewer installations relative to other technology deployments (e.g., the automobile).

Insurers use the "law of large numbers," which says that "the larger the group of units insured, the more accurate the predictions of loss will be" (Ins. Info. Inst. 2009b). Although PV technology has been in existence since the 1950s, deployment of PV at a significant level occurred much later. Between 1998 and 2007, new annual on-grid installations in the non-residential sector increased from 87 to more than 1,500 (Sherwood 2008), as indicated in Figure 1.

Thus, enough solar PV installations have not been in place for long enough for underwriters to feel they can accurately predict what the losses associated with them would be. Because insurers have access to data on only projects that their companies insure, the more insurers there are, the less data each insurer has.

Adding to the complexity of insuring PV installations is the typical involvement of many parties, including installers, developers, investors, lenders, and insurance companies. Each party to the transaction attempts to minimize the risks it assumes while defining the recourse available in case a risk event occurs and leads to actual losses.

In fact and in many cases, insurance products can be used to mitigate and manage the allocations associated with complex contractual arrangements.

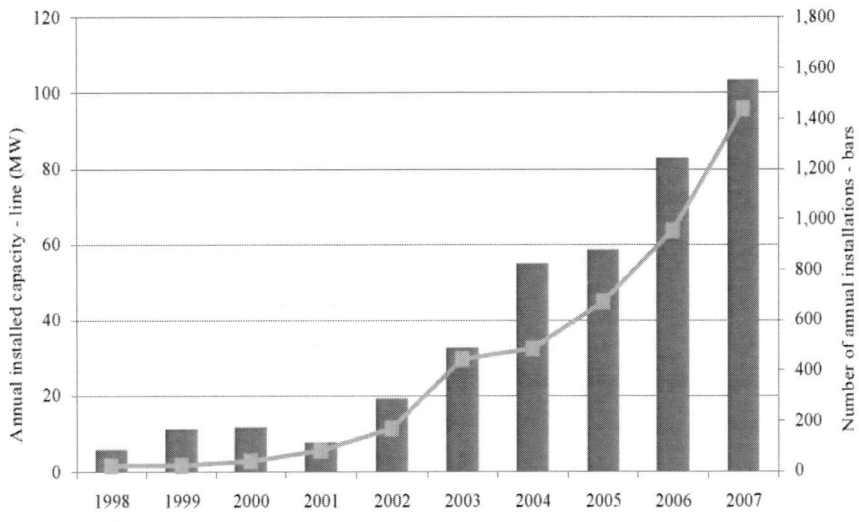

Figure 1. Annual new on-grid installations in the commercial and utility sectors in the United States by numbers installed and capacity installed (Sherwood 2008).

Until the recent growth in the PV market, demand for PV system coverage was not adequate to encourage insurance underwriters to develop PV-specific products. If the upward trend of new PV installations continues, there will be a great demand for insurance products for PV.

This growing market of new solar installations, which is being driven by state policies, federal incentives, and corporate responsibility, represents a possible market opportunity for insurance underwriters.

The focus of this report is commercial-scale and utility-scale PV systems as they represent most of the new market for insurance products. Utility-scale PV projects represent a fledgling but growing market as indicated by the number of proposed projects, many of which have long-term power purchase agreements (PPAs) for power generated. Residential PV installations do not draw attention from insurance companies because they can usually be included under a homeowner's policy (DOE 2003) as long as they are rooftop-mounted.[1]

Interviews with developers of both commercial- and utility-scale PV projects indicate that acquiring insurance at a reasonable cost is a continuing concern for both developers and their customers.

2. GOALS AND METHODOLOGY

Solar technologies are still developing. The insurance industry considers the understanding of the technology and the operational performance of PV systems to be still evolving. A primary goal of this report is to identify financial and informational barriers that developers encounter when insuring solar PV systems. Another goal is to examine information challenges that insurance underwriters and brokers confront when managing risk for PV systems. From these investigations, this report also provides suggestions to the U.S. Department of Energy (DOE) program managers and staff on how these information gaps could be addressed.

Ancillary goals for this report are to (1) help developers and system owners better understand risk management products and the underwriting process for solar PV systems, and (2) inform end users and DOE program managers on how insurance costs can affect solar PV deployment. This report also provides renewable energy researchers and policy makers with information regarding insuring solar PV.

Most of the information for this report was gathered through interviews with experts from the relevant industries. In all, NREL interviewed 26 industry professionals, including those representing:

- four underwriting companies,
- three insurance brokerages,
- seven solar development companies,
- one risk modeling company,
- one electronic PV monitoring company, and
- six representatives from solar power industry associations, research centers, and news organizations.

Because some of the information gathered is viewed as proprietary or sensitive, the identities of the companies and interviewees will not be tied to specific ideas presented in the report. Instead, a list of contacts of those who granted permission to be included in the report can be found in Appendix A.

The interviews consisted of telephone and in-person conversations from November 2008 through June 2009. A complete list of the initial set of questions addressed during the conversations is contained in Appendix B. Follow-up correspondences were also conducted, and all participants were given the opportunity to comment on an earlier draft of the report.

3. THE U.S. INSURANCE INDUSTRY

Underwriters, brokers, and re-insurers all play important roles within the insurance industry and are vital to the U.S. economy. However, the industry is being significantly affected by the current economic downturn, new regulation requirements, and increased losses to policyholders. In the context of these large-scale forces shaping the insurance industry as a whole, the insurance market for PV systems is evolving and maturing rapidly.

The insurance industry is a large part of the U.S. economy. From sales of final goods and services in 2006, the industry produced revenue of $281 billion or more than 2% of the total national GDP ($13 trillion in 2006) (Ins. Info. Inst. 2009c). In all, there were 2,723 property/casualty companies in 2007 and 1,190 liability/health underwriters (Ins. Info. Inst. 2009d). However, the industry has recently experienced major changes that may alter its dynamics.

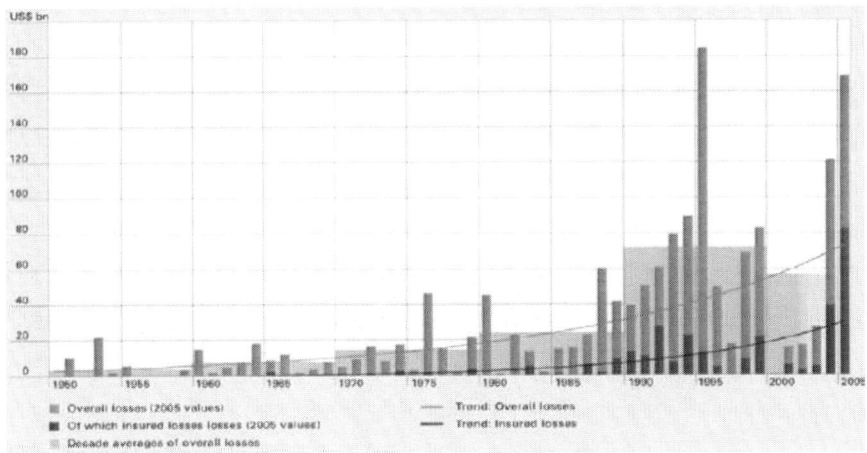

Figure 2. Overall losses and insured losses in the industry: Absolute values and long-term trends (Munich Re 2005).

Industry revenues have declined significantly in the past year, leading to an average annualized rate of return of 1.1% in the first nine months of 2008, down from 13.1% during the same period in 2007 (Hartwig 2008). This sharp decline in revenues is attributed to the financial and economic crises of late 2008, which negatively affected even the most conservative portfolios (Ins. Info. Inst. 2009d). The economic downturn is expected to lead to additional bankruptcies and concentration in the industry. Financial losses in the insurance industry resulting from natural disasters appear to be increasing. As Figure 2 indicates, total losses (and the subset of insured losses) have increased significantly since 1960. In fact, insured losses rose from negligible amounts during the mid-century to around $30 billion in 2005. Uninsured losses increased from around $5 billion in 1950 to well over $70 billion in 2005. Figure 2 also shows that the spikes in insurance losses since 1990 are greater than earlier ones. Figure 3 illustrates the number and type of natural catastrophic events that have occurred over the latter half of the 21st century through 2005. Some insurers are concerned about the link between greenhouse gas emissions and climate change and the potential for increased insurance losses. Insurance industry leaders, such as Munich Re, concluded in a recent report that increased natural disasters, specifically hurricanes, coincide with increased global temperatures, which also coincide with increased greenhouse gas emissions (Munich Re 2005).

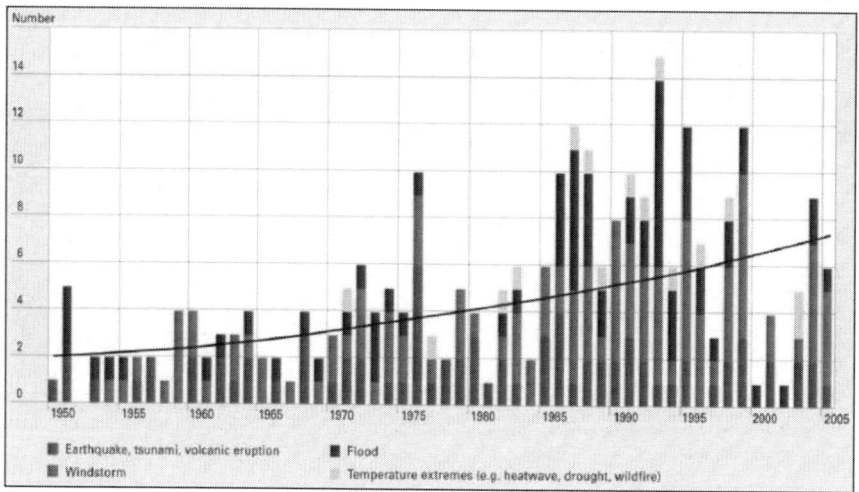

Figure 3. Annual number of great natural catastrophes by event type[2] (Munich Re 2005).

Starting in 2010, new regulations by the National Association of Insurance Commissioners (NAIC)[3] require insurance companies with annual premiums of $500 million or more to disclose all financial risks associated with climate change as well as any plans for mitigating exposure to these risks (NAIC 2009). NAIC put this requirement into place so that insurers would disclose how climate change risk would affect their portfolios and how they plan to change their investment strategy. This disclosure provides insurance underwriters with an additional financial incentive to assess climate change-related risks in their portfolios as their underwriting practices could be scrutinized by concerned investors.

Insurance underwriters are the companies that pay the insured when claims are made. Underwriters formulate the payment estimates designed to cover their risk, which is outlined in a binding contract. Underwriters define the price and risk allocation terms under these contracts through a multitude of processes, including research, engineering analysis, risk modeling, contractual negotiations, and modification of pre-existing underwriting forms, which are described in Figure 4. Insurance underwriters tend to specialize in either property- or liability-oriented policies. According to one interviewee, specialization allows an underwriter to gain expertise in a particular area, which leads to a better understanding of the risks involved. In turn, this might allow customers to receive lower insurance premiums and faster turnaround on the creation of an insurance policy. Nonetheless, some underwriters will write policies for both liability and property, depending on the project.

Within the type of insurance they provide, most insurance companies' products are generic and homogenous, and therefore are not easily adapted to provide protection for renewable energy technologies. However, industry-specialist brokers work with renewable energy product manufacturers, facility developers, and project finance clients to create unique product offerings and the necessary underwriting support that are not well covered by traditional insurance.

The goal of an industry-specialist or consultative insurance broker is to strategically partner with clients and provide innovative, responsive, and cost-effective solutions that mitigate the risk and uncertainty of renewable energy systems. The process of establishing insurance policies begins with developers or project owners identifying the risks that they need or want to cover. This can be done with the help of an insurance broker who might have suggestions for what to cover.

The broker then approaches underwriting companies with an outline of the insurance products that interests its client. The client is presented with the range of policies, premiums, and terms of coverage from which to choose. Once the customer decides which policy it wants, the broker returns to the underwriter to complete the policy formation process.

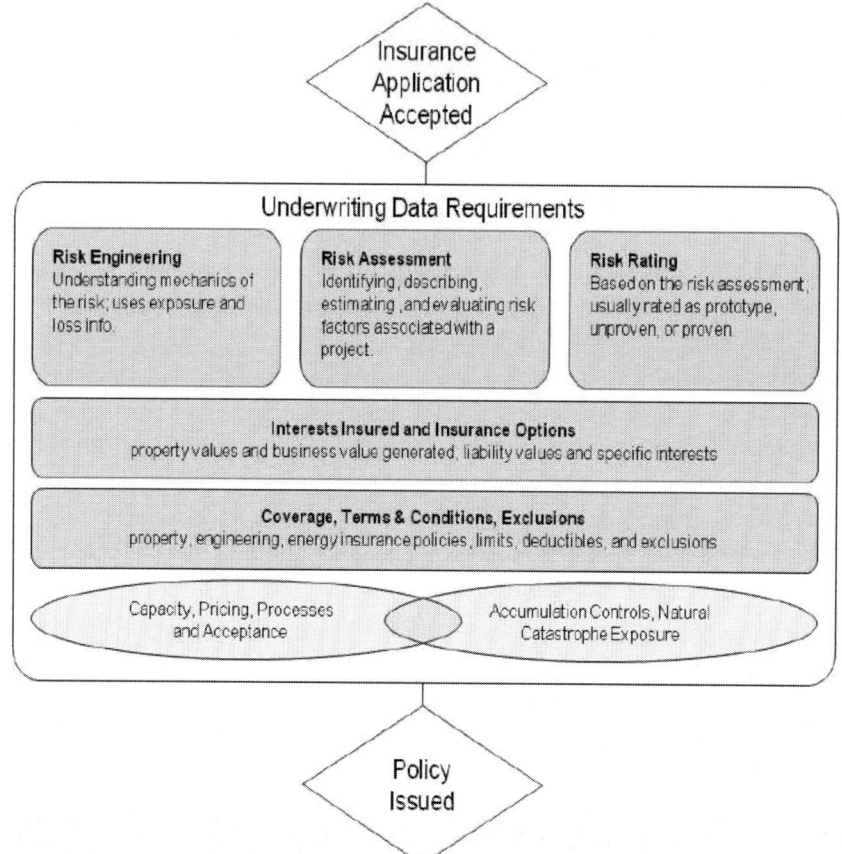

Figure 4. Insurance underwriting process.

Several U.S.-based companies underwrite policies specific to solar PV, a couple of which receive underwriting capacity from Lloyds of London. The role of these companies, which are known as managing general agents (MGAs), is to formulate the insurance policy and sell portions of it to various investors.

The top ten major U.S. insurance companies that underwrite commercial policies (Ins. Info. Inst. 2009e) are listed in Table 1. Insurance products for renewable energy are mostly offered by large insurance companies that provide other commercial and residential insurance products. Of the top ten commercial underwriters, NREL research for this report identified at least six companies that actively write renewable energy policies (Munich Re 2009, AIG 2009a; Zurich 2008; Chubb 2008; Ace Limited 2009; Manning 2009):

- Munich Re (a reinsurer that typically backs the main insurance companies listed above)
- Chartis (formerly AIG)
- Zurich Insurance Group
- The Hartford Financial Services Group
- ACE Limited
- Chubb Group of Insurance Companies

However, it is unclear how the financial crisis might affect the insurance industry and their business practices with respect to renewable projects. Thus, the companies that are able or choose to provide insurance for renewable energy technologies may change.

The European and Japanese solar markets are more mature than the U.S. solar market, as indicated in Figure 5. Most of the growth in PV capacity in Europe has occurred in Germany, which was the world leader in cumulative installed capacity with 5,367 MW in 2008. However, Spain and Japan also have robust PV markets with nearly 3,326 MW and 2,176 MW installed in 2008 respectively. The new PV capacity in Spain and Japan is noteworthy when compared to the United States, which has only around 1,106 MW of installed capacity as of 2008.

The maturity of the European PV market gives European insurance companies more experience underwriting renewable energy generation systems than the U.S. market. Because of Europe's greater experience with solar PV installations, some U.S. insurance companies use loss data from Europe to project probabilities of future losses and risk in the United States.

During the interviews conducted for this research, insurance underwriters raised concerns about the maturity of PV technologies. Other concerns they raised had to do with the lack of widely accepted certification or regulation among installers. Underwriters indicated that owners and developers do not have adequate knowledge about all the risks associated with PV systems. However, underwriters also expressed that the solar PV industry has excellent risk fundamentals and is maturing. These positive insurance and risk fundamentals include:

- Bundled and pooled small-scale projects
- Passive systems with no or few moving parts
- Modularity of construction
- Improved electronic performance diagnostics.

Table 1. Top Ten Writers of U.S. Commercial Lines Insurance by Direct Premiums Written, 2008[a]

Name of Insurance Group	Market Share (%) 2008	Market Share (%) 2007	2-year Growth (CAGR)[b]	Direct Premiums Written (in $ billions) 2008	Direct Premiums Written (in $ billions) 2007
Chartisc	9.03%	10.64%	-9.78%	$20.2	$24.7
Liberty Mutual Group	6.57%	5.45%	13.36%	$14.7	$12.6
Travelers Group	6.26%	6.19%	-1.42%	$14.0	$14.4
Zurich Financial Services	6.04%	6.29%	-3.72%	$13.5	$14.6
Hartford Financial Services Group	3.11%	3.15%	$0	$6.9	$7.3
CNA	3.10%	3.22%	-6.09%	$6.9	$7.4
ACE Group	3.10%	3.12%	-1.43%	$6.9	$7.2
Chubb Group	2.91%	2.88%	-2.60%	$6.5	$6.7
Nationwide Mutual Insurance Co.	2.35%	2.36%	$0	$5.2	$5.4
Allianz	2.12%	1.95%	3.86%	$4.7	$4.5
TOTAL	200844.59%	200745.25%	--	$2107.50	$2111.80

[a] Source: National Association of Insurance Commissioners (NAIC) Annual Statement Database, via Highline Data, LLC. Copyrighted information. No portion of this work may be copied or redistributed without written permission of Highline Data, LLC.
[b] Compound annual growth rate
[c] Formerly known as AIU Holdings and American International Group/AIG.

During interviews, insurance industry experts offered the following view of the PV industry.

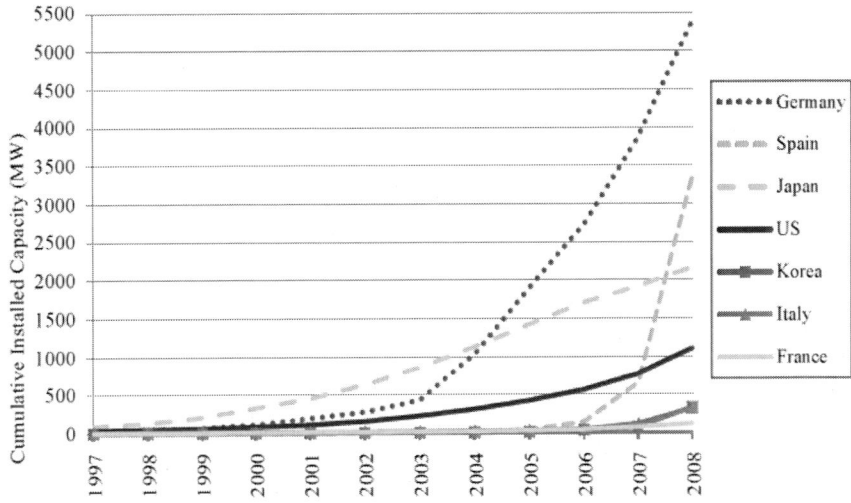

Figure 5. Cumulative installed PV in top seven countries (IEA 2008, REN21 2009, EurObserv'ER 2009, Sherwood 2008).

Insurance Industry View of the Solar Energy Industry

Developers are not educated about the coverage they need.

If the broker does not understand solar PV technology, ensuring that the project is properly covered can be problematic.

Because developers focus on project development, not insurance, they may not know what coverage they need.

Most developers think they pay too much for insurance. However, if they have not fully covered themselves, they may be paying less than they need to pay to cover all possible risks.

The solar PV market is a maturing industry, but it is not yet mature (even for crystalline silicon).

The solar PV industry has excellent fundamentals (e.g., strong product demand, declining input costs).

Similarly, solar PV industry experts also feel that there is a lack of appropriate information among the insurance industry.

> **Developers View of the Insurance Industry**
>
> Insurance premiums for solar PV systems are too high.
>
> Insurance companies sometimes lack the background knowledge of solar PV technologies.
>
> Many do not understand how the technology works.
>
> For a PV system, one insurer asked about the use of molten salt, which is relevant for solar-CSP technology but not for PV technology. Technology information is needed.
>
> If the insurance industry had better data about system operation and historical losses, insurance premiums could go down.
>
> Insurance brokers are considered the "800 pound gorilla" in the room.
>
> Brokers represent the most common way for renewable energy insurance policies to be instigated with underwriters.
>
> If brokers are not educated on the technologies and the risks, then the underwriters will not be educated either.
>
> Some brokers pretend they understand solar technologies and place policies that do not fully cover what needs to be covered. This is an issue for the brokers who do not understand the true system risks.

For example, the PV industry thinks that insurance companies' rates for PV policies are not affordably priced; that the industry is lacking in information about PV technologies; and that insurance brokers may or may not fully understand PV technologies.

Developers offered the following view of the insurance industry during interviews.

Some form of insurance is essential to deploying solar energy projects. Investors look for a structured insurance program that transfers identified project risk to a third-party insurance company. Lenders look to the project and its revenue stream as the collateral asset being financed in their loan underwriting.

However, PV system owners may find obtaining affordable insurance difficult for several reasons. First, only a small subset of the insurance industry insures renewable energy projects. Interviewees indicated that until recently only Lloyds of London underwrote insurance for solar projects in the United States. Reduced competition between large insurance companies likely leads to higher insurance premiums. The following descriptions of the market for solar-specific policies were given by insurance industry experts during interviews.

> **Insurance Industry View of Insurance**
>
> More insurance companies are seeing the opportunity of a new market.
> Some companies are leaders in providing solar-specific projects; others are waiting. The more opportunities companies have to understand the technologies and see actual operational data, the more likely competition is to occur and the more likely premiums will go down.

The insurance industry, including insurance brokers and underwriters, is a large and important one in the United States. However, it is experiencing a number of challenges, such as the financial crisis and increased losses from natural disasters that could alter its dynamics.

Importantly, U.S. insurance regulation will require climate-change risk disclosure starting in 2010. In addition, the industry is developing new risk management products for the maturing PV industry, which will create information uncertainties for both the insurance and PV industries.

The following sections describe the insurance products available for PV, as well as the information the insurance industry needs to develop products that advance PV technologies.

4. SOLAR PV INSURANCE

To remain feasible in terms of risk, non-residential PV installations require a variety of insurance products,[4] including general liability, property, and environmental risk insurance. By purchasing these types of coverage, developers create a financial backstop for the project, without which it would be difficult for the developers to obtain favorable financing.

Most large PV systems require liability and property insurance, and many developers may opt to add policies such as environmental risk insurance (see Appendix C for a full list of PV testing and certification laboratories). The following section discusses insurance products from a high level; however, it is important to note that policies may vary from underwriter to underwriter and on a project-by-project basis.

4.1.1. General Liability Insurance

General liability covers policyholders for death or injury to persons or damage to property owned by third parties. Rooftop installations typically require additional liability insurance given the risks inherent in working on roofs and the higher likelihood of wind loading. Ground-mounted systems tend to be far from other structures and in less-populated areas, which may reduce the premiums for general liability insurance or may reduce the requirement for additional insurance.

General liability coverage is especially important for installers, as risk is greatest during installation. However, solar power generation system owners may also purchase builders' risk insurance in addition to general liability coverage to indemnify themselves from damage to other property or persons during the construction phase.

4.1.2. Property Risk Insurance

Property risk insurance covers "damage to or loss of policyholders' property" (Ins. Info. Inst. 2009f). While the manufacturer's warranty will provide some limited defect coverage, the system owner usually purchases property insurance to protect against risks not covered by the warranty or to extend the coverage period. Property insurance also protects the owner against financial loss from theft of system components, which insurance underwriters and brokers consistently mentioned as a concern, especially before the panels are affixed during construction. In addition, property insurance can indemnify system owners of certain natural catastrophe risk, which—according to one insurance underwriter—is the second largest risk component of property coverage after the risk of theft.[5] If natural catastrophe risk is perceived to be too high, separate policies may be needed to provide additional risk coverage capacity. Examples of additional policies for location-specific natural catastrophe risks include:

- Earthquake coverage in California
- Hurricane coverage in Florida
- High-wind coverage in Colorado.

Property risk insurance also covers the transit of goods, such as material shipped to the project site, particularly for modules and components that are manufactured internationally and are needed during the construction phase or are later returned to the manufacturer for repair.

4.1.3. Environmental Risk Insurance

Environmental damage coverage indemnifies system owners of the risk of either environmental damage done by their development or preexisting damage on the development site. There are a variety of environmental policies that can cover an assortment of risks (Jones 2001), such as:

- **Pollution legal liability policies** cover the insured from risk with unknown pollution conditions as well as liability for harm caused to persons by the pollution. This type of policy also covers business interruption and transportation claims, but it does not cover the cost of long-term cleanup efforts (Jones 2001).
- **Property transfer policies** transfer the risk from the insured for existing pollution claims and pays for claims under terms similar to Pollution Legal Liability (Jones 2001).
- **Cleanup cost cap or stop loss policies** are customized policies that create a cost stopgap for continued cleanup efforts or for newly found contamination as well as bodily injury. This type of policy does not usually cover property damage or legal costs (Jones 2001).
- **Brownfields restoration and redevelopment policies** indemnify policy owners who are developing projects on sites that are known to be contaminated from ongoing high costs, bodily harm, legal costs, costs for cleaning up of unknown additional pollutants, and property damage (Jones 2001).

Only a few of the major U.S. underwriters, including Chartis (formerly AIU Holdings/AIG), Zurich, and some smaller entities, offer environmental risk coverage (AIG 2009b).

4.1.4. Business Interruption Insurance

Business interruption insurance is often required to protect the cash flow of the project. This coverage ensures that policyholders can recover:

1) Lost sales as a result of the system not being operational and loss of production-based incentives also resulting from the lack of electricity production
2) Recapture of tax incentives lost because of the project not being rebuilt or not being rebuilt quickly enough.[6]

Projects financed under third-party ownership structures (described in Section 5.1) generally require the procurement of business interruption coverage.

4.1.5. Contractor Bonding and Construction Risk Management

Construction of PV and other renewable energy facilities entails unique risk properties and solutions. Because of an array of risks related to performance and safety, contractors and sub-contractors are generally required to be bonded (i.e., hold a surety bond to cover liens held for poor performance or misappropriated funds). Banks and insurance agencies provide contractor bonding. However, because of the minimal track record for developing renewable energy systems, all but the largest contractors are often unable to obtain bonding. Project lenders almost universally require that all contractors and subcontractors be fully bonded relative to the value of work to be completed. Without adequate bonding, contractors may not participate in project development, thus lowering competition for contractor services.

One company, Broadlands Financial Group, LLC (Broadlands), offers a unique alternative to bonding requirements. Services similar to Broadlands' (referred to as construction risk management) negate the need for contractor bonding. Construction risk management is a series of due diligence, system performance, and funds control protocols designed to assess and off-load various risks related to contractor performance. Broadlands' services are offered by a number of competing firms for non-renewable energy projects but—according to the company—Broadlands is the only provider of third-party construction risk management services for renewable energy projects. The company's first renewable energy project was the 64 MW Nevada Solar One facility where they managed $120 million in construction and more than 100 suppliers and contractors (Broadlands 2009).

Construction risk managers, such as Broadlands, are generally engaged by the lender. The services offered by a construction risk manager include, but are not limited to:

- Reviewing plans and ensuring bid sufficiency
- Assessing contractor financial qualifications (check that the contractor is not overextended)
- Assessing bids to ensure proposed costs are consistent with generally accepted levels
- Establishing bank account similar to an escrow
- Verifying that the work invoiced was completed
-
- Ensuring funds are properly disbursed to sub-contractors, if relevant
- Coordinating lien releases.

According to Broadlands, construction risk management can save up to half the cost of surety bonds, which represent approximately 3.0% of total construction costs. Broadlands guarantees system performance for facilities for which it manages construction risk. And according to the company, the services are available to small installations of approximately $0.5 million in total development costs.[7]

Property insurance typically covers system components beyond the terms of the manufacturer's warranty. For example, if a PV module fails for reasons covered by and during the manufacturer's warranty, the manufacturer is responsible for replacing it, not the insurer. However, if the module fails for a reason not accounted for in the warranty, or if the failure is beyond the warranty period, the insurer must provide compensation for the replacement of the module. In this way, warranties can have a positive impact on insurance premiums as they indemnify the PV system owner in the event a product defect leads to losses.

Although very rare, manufacturers can also seek risk coverage from insurance companies for serial defects on their products. A manufacturer can be held accountable for replacing all products associated with a serial defect. This was the experience of several wind turbine manufacturers during the early 1990s that were unable to maintain adequate quality controls during a period of very high demand. Also, recent examples of massive recalls involved Suzlon wind blades and REC Solar, Inc. solar panels (Greentech Media 2009, CompositesWorld.com 2008). According to one underwriter, warranties can be guaranteed for up to ten years. However, underwriters limit the capacity they will provide to a single manufacturer for serial defects on a specific PV module or component. This type of coverage is therefore specialized and expensive, but it is still available and used in the market. Munich Re, Marsh Insurance Brokers, and Signet Solar recently announced the development of a new insurance product that guarantees the performance of Signet's solar PV modules for 25 years (Signet, Munich Re, and Marsh 2009). This warranty guarantees performance up to 90% of capacity for the initial 10 years and 80% of capacity for the remaining 15 years, which provides greater business certainty for project developers and is expected to lead to more favorable financing and better project economics (Signet, Munich Re, and Marsh 2009).

5. INSURANCE ISSUES SPECIFIC TO THE THIRD-PARTY OWNERSHIP MODEL

To take advantage of federal financial incentives,[8] developers have structured solar power projects to enable third parties to invest equity and take advantage of the tax benefits. In such a case, the interested third party has "tax appetite" or taxable income that it desires to shelter. And, the solar project developer (potentially the same entity as the equity investor) acts as an intermediary with the customer, providing critical services such as system design, installation, contractual arrangement, and system maintenance. The contract between the project developer and the customer regarding the procurement of the system's power takes the form of a power purchase agreement (PPA) or, as offered more recently, an operating lease to private entities. This financial and development structure, which is generally referred to as the third-party ownership model, has increasingly become a common, cost-effective means of developing PV and other renewable energy projects.

Because of the increasing importance of the third-party ownership/PPA model, NREL formulated specific questions for interviews held with PV project developers, insurance brokers, and underwriters about the complexities of obtaining affordable insurance products for projects structured this way. According to the interviews conducted, the price and difficulty of obtaining insurance is of significant concern when developing solar PV projects using this model. And some interviewees claimed that the price of insurance has had a "dramatic impact" on the feasibility of certain PV installations and has been high enough to stop development of some facilities. However, underwriters interviewed pointed out that property insurance is directly tied to property values. Therefore, property coverage costs will decrease as PV costs decrease. Also, PV installations have no fuel costs, low O&M costs, and low taxes because of accelerated depreciation. Thus, insurance costs may appear to be high in relation to other operating costs, although they are more reasonable when compared to the total installed cost.

The following section explains the structure of the third-party ownership/PPA model and the particular challenges it poses for third-party owners and developers.

The third-party ownership/PPA model allows a private or public entity to host a system on its property; another entity that can take advantage of the tax benefits (or treasury grant) owns the system.

The host enters into a long-term contract (the PPA) with a third party to purchase the electricity generated on its property. The electricity price is typically set at a level competitive with the host's retail rate for the first year and then typically increases at a fixed rate over time.

The developer manages all aspects of the system (financing, installation, and maintenance) and bears all operating risks.

The developer also monetizes the environmental attributes of the PV system by separately selling the renewable energy certificates (RECs) to an electric utility (generally, the investor-owned utility that serves the customer), allowing the utility to meet its renewable portfolio standard (RPS) as established by the regulating public utility commission.

In most cases, the third-party ownership/PPA model incorporates an infusion of equity capital from an external entity seeking to reduce its taxable profits.

The details of the roles and responsibilities of different parties for one variation of the third-party ownership/PPA model are shown in Figure 6. Benefits of the third-party ownership/PPA structure include (Cory et al. 2009):

- **The ability to monetize federal tax incentives** (ITC and MACRS), through a third-party project investor, lowering the overall cost to the host entity
- **Low/no up-front costs** for the site host. Instead, the up-front costs are transferred to the developer and project investors.
- **A pre-determined electricity price for term of contract,** for the portion of load served by the PV system. This typically approximates the customer's current retail rate in the first year and then usually escalates annually.
- **A shift of operations and maintenance responsibilities** to a qualified third-party project developer
- **A path to PV system ownership.** If negotiated as an option in the PPA, the host entity can usually purchase the PV system at some time after year six.[9]

Before 2008 was over, Greentech Media estimated that 65% to 70% of the commercial market in that year would use a PPA for PV installations, an increase from previous levels of 50% in 2007 and 10% in 2006 (Greentech Media 2008).

Because the third-party ownership/PPA model now represents the majority of commercial and industrial installations, insurance issues relevant to this financial structure are likely to be an important factor for continued development of the solar PV industry.

Developers interviewed who use a third party/PPA financing model indicated they generally develop systems 100 kW or larger; however, one developer seeks projects with a minimum size of 500 kW. Developers tend to focus on medium- to large-scale projects because smaller installations suffer from a lack of economies of scale, and thus are difficult to develop cost effectively.

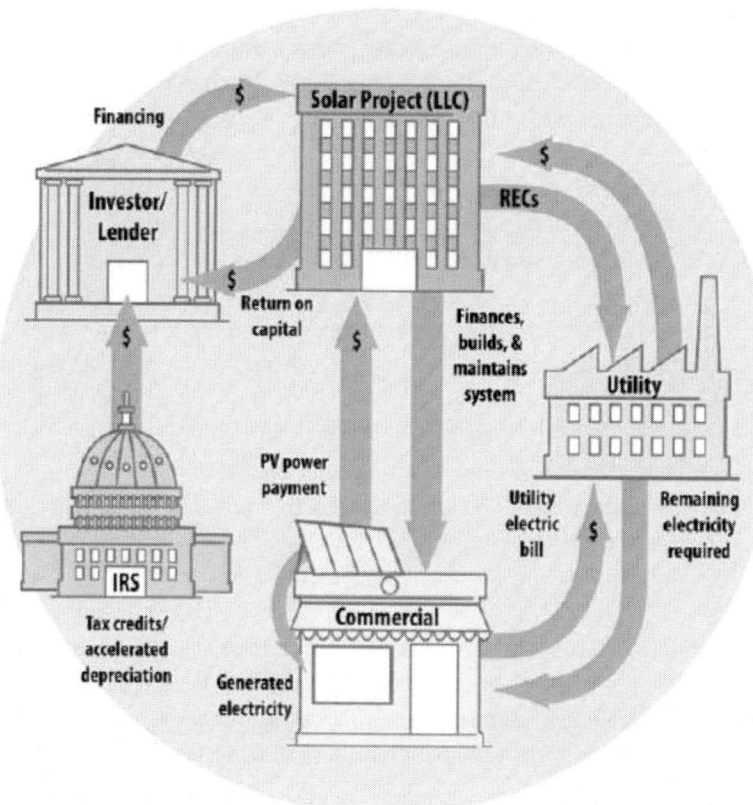

Figure 6. Contracts and cash flow in the third-party ownership/PPA model.

The lack of economies of scale is relevant to smaller projects for reasons beyond insurance, including installation costs, legal fees, maintenance, and other factors. Some developers seek out "structured policies" that are volume-driven (i.e., multiple installations) or scalable to the size of the project.

Under a PPA structure, the unique risks and associated insurance needs are allocated among the various participants. Generally, the developer needs to procure the majority of the insurance products, but each developer may prefer to structure projects in a unique manner.

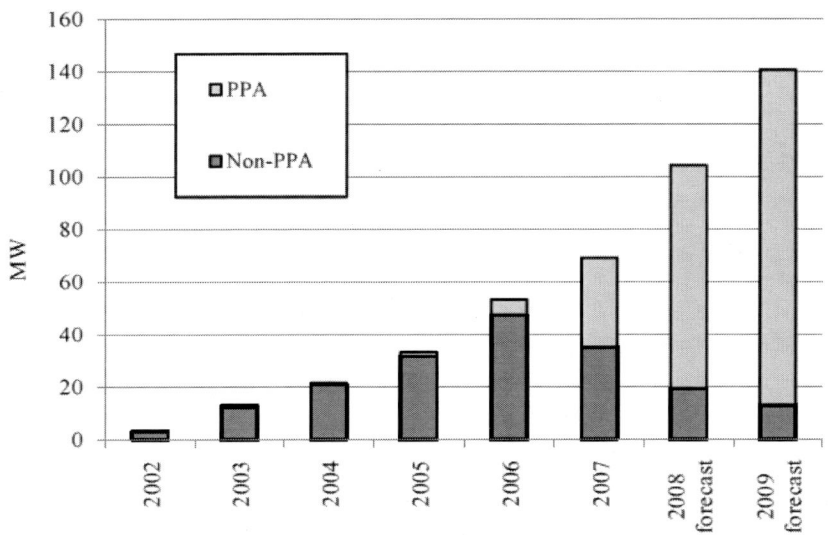

Figure 7. Annual commercial PPA PV installations (Greentech Media 2008).

5.2.1. System Owners

The primary insurance requirements for project developers are property insurance and general liability insurance. Some projects also acquire environmental insurance if environmental issues are associated with the facility site. Separate insurance products are required for the construction period and the operating period. Property insurance protects the owner's investment in the system itself in case of damage to the insured property. Liability insurance protects against financial losses that result when an insured property damages other property or people. Environmental insurance protects both against environmental damage and personal harm caused by pollution to the property done during development and extant damage discovered during the construction of the project.

The interviews with industry experts revealed that location-specific conditions also affect insurance for PV projects. Certain insurance policy riders, which are often based on geographic considerations, can cost too much and can significantly limit new solar project development. For example, an interviewee from a large solar development company indicated that they do not attempt to develop solar projects in Florida because of the high cost of hurricane-related insurance. Other developers stated that because of price, they forego earthquake insurance in California and hurricane insurance in Florida and "self-insure" against these risks (i.e., do nothing and hope that nothing happens). Others procure separate earthquake or hurricane insurance—as required by third-party investors—and pass the cost to the site host via higher power prices or additional terms of the PPA. Developers that procure projects in these states often reduce their overall exposure by diversifying the locations of their solar installations.

5.2.2. Land Owners/Site Hosts

Depending on the project, specific risks are borne by the property owner. One developer interviewee indicated that for its projects, property owners are only responsible for business interruption insurance. This insurance product generally covers up to one year of business income due to significant property damage associated with the PV installation. Another developer interviewee indicated that it requires all customers (residential and commercial) to incorporate the PV installation into an existing property insurance policy. According to that developer, customers can obtain better quotes from the insurance market and thus lower overall project costs. This can be especially true for federal or state agencies.

The third-party ownership financing model is prevalent in the solar industry today, and its dominance is expected to continue. Therefore, developers were specifically asked how insurance requirements and premiums have impacted this financing structure.

5.3.1. Risk Assessment under Third-Party Ownership

Different contract structures applying the third party ownership/PPA model allocate the array of risks to the associated parties in different ways. One underwriter indicated that the strength of the contract between the system host, the developer, and the tax equity investor is very important for determining the types and degrees of risks involved. Some underwriters indicated that they prefer to insure projects that use the third-party PPA model because they view the contract as a positive risk modifier. Other underwriters noted that the contracts vary greatly in content, which can complicate their risk assessment of the project.

5.3.2. Conflict between Long-Term PPAs and Short-Term Insurance Premiums

The third-party ownership model presents a particular challenge to developers when they try to estimate PPA prices for their customers. Insurance products are generally offered for one-year periods, while PPAs are usually offered for 15 to 20 years. Developers cannot determine the cost of insurance two years out, let alone 20. Also, insurance rates can rise on an annual basis during project construction or more likely during project operation. They typically increase to take into account perceived increased risk and inflation. Of course, the increased cost of insurance over time negatively affects project economics. Under these conditions, developers might find it difficult to competitively price their projects to potential customers.

5.3.3. Range of Insurance Costs

Developers estimate the annual cost of insurance to be around 0.25% of the total installed cost of the project, which could be as high as 0.5% annually in areas where extreme weather events are likely. While this sounds quite low, the annual costs do add up over a 20-year contract, especially since annual premiums escalate, usually every year. In addition, insurance premiums take up a sizable component of the operating budget. One developer indicated that insurance is their biggest operating cost. Another developer indicated the cost of insurance is 25% of annual non-capital costs. While this is less than their annual cost of administration (42%) and operation and maintenance (33%), it is still significant in the context of operating costs for solar PV installations. Although insurance is not the largest operating cost component, it is considered the most uncertain, mainly because the annual premiums can increase significantly if more/larger catastrophic events occur in the region or state than expected. The fact that the developer must estimate these costs up front means that they might hedge themselves to cover the potential costs for the entire 20-year period, based on current-year insurance prices and a significant insurance premium risk escalator. Property insurance represents the majority of the overall cost of insurance. One interviewee indicated property insurance represents roughly 90% of the total cost of insurance. Another developer indicated 96% of the total insurance budget is allocated to property insurance. According to one developer, insurance costs 5% to 10% of the total delivered price of energy from the PV installation. The same developer also indicated that lower insurance premiums would make it much easier to increase market penetration. Another developer indicated a low quote for insurance can give a competitive edge when bidding on a project. It was also noted that, historically, a single insurance provider offered coverage for a given project. More bids and better prices have been available to recent projects. Finally, a developer noted that global insurance companies offer better prices than American insurance companies, which have not been as aggressive in trying to underwrite the renewable energy market.

6. INSURANCE CHALLENGES FOR RENEWABLE ENERGY PROJECTS

Obtaining affordable insurance can be challenging for the fledgling renewable energy industry, partly because insurance brokers and underwriters are not as familiar with renewable energy technologies and the associated risks as they are with other technologies they underwrite. Accordingly, only a small niche of the insurance industry serves the renewable energy market. This section explores the specific challenges that solar PV faces when securing insurance.

In our interviews, insurance brokers and individual insurance companies consistently viewed insuring PV technologies as an emerging niche market as neither the technical aspects of photovoltaic modules nor the physical infrastructure necessary to support PV – on rooftops or in ground-mount configuration – is widely understood. Moreover, the solar PV market is booming with new producers and technology. While this boom encourages technological innovation, it complicates underwriting. Many insurers do not have the "risk appetite" (i.e., do not feel it is economic to assume this risk) for PV products that they consider "prototypes." The definition of "prototype" varies depending on the insurer. Definitions of prototype vary from a solar PV system that has been operating for less than a year to a technology that has not yet been commercially manufactured. The following lists some views from insurance underwriters and brokers on the definition of a prototype that were offered during interviews.

Although there is not a strict definition of PV product prototype, underwriters are clearly uncomfortable with accepting the risks associated with prototypical technologies at normal premium rates. Several underwriters indicated that to understand fully the risks associated with an installation, they need high-quality data on specific modules presented by company and module number. Thus, data on the performance of certain technologies or various products produced by a company would not help underwriters of different products in their risk assessments.

Often, test data are for "bench-scale" facilities and not full-scale installations. However, developing full-size test installations for every module design can be expensive for developers or manufacturers. In Section 7, we discuss potential opportunities for NREL and DOE to facilitate improved testing capabilities for new module design. Additionally, once a solar technology goes beyond being a prototype (at least a year of operation), one or more additional years may be required for the technology to be considered "proven." Nonetheless, even time series data for less than a year would be helpful, according to underwriters interviewed for this analysis.

Insurance Industry Definitions of Solar PV Prototype

Fewer than 3,000 total hours of actual time in operation per manufacturer's product (e.g., Sharp PV panel model; First Solar PV panel model #)

A range of 3,000 total hours of actual time in operation (for PV and technologies without moving parts) to 8,000 total hours of actual time in operation (for wind and natural gas power plants, i.e., anything with moving parts)

Any technology that is commercially unproven and has not been deployed widely

A technology that is not commercially deployed even if it has one or more full-scale examples installed

Any product with fewer than 8760 hours or one year's worth of operational data

---

Risk assessment and pricing requires access to and statistical evaluation of large quantities of verifiable data; the industry uses its analysis of various data (e.g., operational and loss history data) to calculate the probabilities of property, liability, environmental, and other losses. As a nascent industry, the solar industry has relatively little data with which it can assess risks. Some insurers with subsidiary or sister companies in Europe can apply their historical loss data to U.S. applications, assuming they are applicable under similar circumstances. U.S.-based companies without access to such information may be unwilling to pursue renewable energy projects. The absence of loss history data limits the ability of insurance underwriters to assess clearly the risks of insuring solar PV systems. As a conservative industry, the insurers will therefore add a risk premium so that they can be sure to cover any potential risks. Greater access to data will allow the insurance industry to assess more accurately the risks associated with PV facility development and operation and will lead to more accurate insurance premiums.

Interviewees indicated entities in the renewable energy industry are not properly delineated in existing business classification protocol. For example, manufacturers are classified via Standard Industrial Classification (SIC) and

North American Industry Classification System (NAICS) codes, which should be unique to manufacturers of renewable energy components and systems, according to interviewees. Additional delineation—perhaps through an incremental coding system—could allow insurance underwriters to better assess insurance claims related to workers' compensation claims and the associated operating loss data for the renewable energy industry. Interviewees made similar comments with respect to renewable energy project installation, development, and ownership. Insurance industry professionals described generally poor comprehension of entities' business models and the associated risks, which they attributed to poor data collection classifications and risk assessments specific to the renewable energy industry.

Several interviewees indicated the need for more testing of modules prior to commercialization and better comprehension of risks through scientific evaluation. At present, Underwriter Laboratories (UL) and other organizations (including NREL) test modules and inverters for various risks and performance characteristics. However, both insurance and solar industry representatives recognized the need for more evaluation of different stresses on PV systems, including wind, hail, and extreme temperatures as well as the need for improved coordination of standards and procedures in testing. Members of the insurance industry interviewed would like to see the testing data for a full year for each type of module made by each PV manufacturing company.

Several underwriters indicated that catastrophic risk models are an important part of the risk assessment process. A handful of risk modeling companies develop and produce these risk models exclusively for the insurance industry. Underwriters use these tools either to predict risk based on related criteria or historical loss performance.

Risk Management Solutions (RMS), AIR Worldwide, and EQECAT are among the few major providers of catastrophic risk management models who develop industry-leading predictive software. The models employ modular design to assess the type of occupant and exposure to catastrophic risk, including the likelihood of severe weather events such as earthquakes, storm-strength winds, and floods. The RMS model, for example, evaluates risk based on 32 categories of occupancy and predicts occupancy and liability risk based on mathematical computations of risk scenarios (e.g., how likely a Category 3 hurricane is to occur in a given location and cause equipment damage, business interruption, loss of life, and dismemberment).

The RMS model aggregates all rooftop equipment (such as air conditioners and PV panels) into a single category of relevance that represents

RMS's customers for the underwriters. The RMS software also evaluates various manufacturing facilities, including energy production facilities. The model groups energy projects into subsets, including nuclear, fossil, transmission and distribution lines, switchyards, and renewable projects. The model does not distinguish the type of renewable project. The risk models used by the insurance industry could be improved by providing both access to more extensive operational history and better distinctions of rooftop equipment, including design of specific PV modules, installation techniques, and relevant equipment.

As the PV industry has grown quickly, the number of PV installers has also grown. Several developers raised concerns that contractors (e.g., electricians and plumbers) working on PV projects often have no formal solar PV installation training. Although thorough voluntary training is available through the North American Board of Certified Energy Professionals (NABCEP), there is no mandatory certification program for installers of solar PV systems. Both developers and insurance underwriters raised quality and safety concerns regarding the lack of consistent certification of contractors and subcontractors.

7. RECOMMENDATIONS AND SOLUTIONS

A range of policies could be implemented to increase the availability of information, improve the comprehension of the associated risks, and—in time—lead to better availability and lower cost of insurance products.

Specific government policies and assistance can improve the availability of information relevant to the insurance and solar industries. In this section, we explore opportunities to increase the flow of information about solar PV.

7.1.1. Improved Solar Technology Information Dissemination to the Insurance Industry

The insurance industry is—almost by definition—risk averse, and thus slow to expand their coverage into unfamiliar terrain. Because the solar industry is evolving quickly and many of its technologies and components are highly technical, maintaining current understanding of the industry can be difficult. Education of the insurance industry (including the technologies available and the installations and operational track records of facilities) could reduce the perception of risk. Specific educational products and services targeted toward the insurance industry could bridge information gaps and lead to more competition among insurance companies vying for PV underwriting opportunities, which could in turn lead to reduced cost of various insurance products.

7.1.2. Improved Availability of Solar PV Historical Loss Data

The insurance industry, by definition evaluates and quantifies risk, which requires intensive study of data. As the PV industry is still in its infancy, gathering high-quality, long-term data regarding system performance, property loss, liability, and other factors is difficult. One company, Fat Spaniel Technologies, is assembling a database of PV system experience. Fat Spaniel, a corporation that electronically monitors 2,000 mostly larger renewable energy systems around the world, refers to its product as an "asset management application." Aggregated data from such continual monitoring systems could be used to assess historic loss data and associated insurance claims.

However, more could be done to assemble and evaluate historical operating data for PV systems. Specifically, the DOE could play a valuable role in working with industry associations, system installers, and operators to assemble data and evaluate the availability and operating statistics, type of module and inverter technology deployed, geographic location, facility orientation, building type, and installation design. For example, by increasing the aggregate information available to the insurance industry, the resulting database tool could lower perceived risk and reduce insurance premiums. Such a database could also inform the market to implement the most productive solar facilities possible given other relevant criteria.

Even if the industry were not interested in partnering, the DOE would likely be able to gather high-quality performance data from PV installations whose developers qualify and who opt to take the federal Treasury Cash Grant in lieu of the federal ITC. This is a result of the DOE requirement that qualifying PV installations report performance in kilowatt-hours annually for five years (Treasury 2009). While this data may be a starting point, it will not provide information regarding the types of losses incurred at the various installations, which is of most use to insurance underwriters.

7.1.3. Improved Classification of Renewable Energy Entities

Further investigation of the classification of renewable energy component manufacturers and system developers, installers, and owners is warranted. Manufacturing entities are classified by the NAICS, which is organized by the National Technical Information Service (NTIS), a branch of the Department of the Commerce. The last NAICS classification was published in 2007 and is based on information prior to 2005.[10] The NAICS groups numerous industries under each code. For example, the code for alternative energy structure construction (# 237130) also covers cable laying, nuclear power plant construction, and microwave relay tower construction. Solar energy power generation via photovoltaics (# 221119) shares the same code with wind, tidal, and concentrating solar power generation.

SIC codes, which were created in 1937, are administered by the Occupational Safety and Hazard Administration of the Department of Labor. The four-digit system, which was generally replaced by the six-digit NAIC system in 1997, is still used by certain agencies such as the U.S. Securities and Exchange Commission. Insurance entities have difficulty gathering and assessing claim and loss data to properly determine the associated risks and price-relevant insurance products. Work could be done with the U.S. Departments of Commerce and Labor to improve the NAICS and SIC code definitions, which would improve the accessibility and specificity of data that are necessary and relevant to the insurance industry.

7.1.4. Module and Component Testing

When bringing solar modules and other components to market, it is important for manufacturers to certify the capability of their products under a variety of conditions. Testing by the Underwriters Laboratory (UL) is required to ensure safe operation. While the UL test addresses safety, it does not attempt to ensure performance.[11] The California Energy Commission (CEC)[12] and many other U.S. organizations require safety testing to UL 1703 but do not require qualification testing. The International Electrotechnical Commission (IEC) introduced a series of test procedures (IEC 61215 for crystalline silicon (Si) modules, IEC 61646 for thin-film modules, and IEC 62180 for concentrating PV modules) that is gaining recognition as the standard certification procedure. The IEC procedures, which include 18 tests of PV modules and other solar components, apply various stresses to the modules to identify design flaws that could lead to early failures. These tests are consistently required for installations in Europe. An effort is underway to require both safety and qualification testing within the United States to improve the consistency of performance by PV products in the field.

At present, most European organizations require IEC 61215 for any Si-based modules being brought to market. California, which aggressively supports PV deployment, requires UL testing (1703) by a nationally recognized test laboratory and additional performance-parameter testing at an

International Laboratory Accreditation Cooperation (ILAC) affiliated laboratory. Only four U.S. laboratories conduct the testing on PV equipment required by the CEC:

- Florida Solar Energy Center
- Intertek Testing Services NA, Inc. (formerly known as ETL)
- TÜV Rheinland PTL, LLC (formerly known as ASU/PTL)
- UL Photovoltaic Technology Center of Excellence

The complete list of national and international testing laboratories that offer the certification required by the CEC is provided in Appendix C.

NREL conducts a variety of tests on PV modules but does not provide certification services with respect to IEC 61215, IEC 61646, or IEC 62180. NREL's testing capabilities are available to cell and module manufacturers, but funding for such services must be provided by the private sector entities. However, cost recovery from module manufacturers is an overly complex process requiring extensive paperwork.[13] NREL and the DOE are working on improving the certification service charge process to enable NREL to expand the availability of these services. Aside from certification procedures, there is a broader need for testing capabilities of near-market technologies. While certain public-private partnerships develop materials and components, there is no definitive method by which new technologies or technology combinations can be displayed to potential developers in the United States who may be interested in using the products. Fortunately, the Solar Technology Acceleration Center (SolarTAC)—a testing facility offering land and development infrastructure—is in development in Aurora, Colorado. The facility, which is expected to be complete by end of 2010, offers "founding members" five acres of research space, certain site infrastructure, and the opportunity to collaborate with other members. At present, founding membership is limited to Abengoa Solar, SunEdison, and Xcel Energy. NREL is limited in its participation in SolarTAC because of restrictions on paying membership fees using DOE funds. Moreover, SolarTAC's location in Colorado effectively eliminates testing for higher humidity environments, such as the southeastern United States. Beyond the capabilities of SolarTAC, module manufacturers need to showcase their technologies in commercial-scale projects, including by interconnecting to the grid. The financial community generally wants to fund only projects that have been tested in commercial contexts. Even as government incentives (such as Treasury grants and DOE loan guarantees) provide significant financial support to new PV projects, developers and financiers have minimal desire to incorporate untested technologies in projects. NREL and DOE could play an important role by developing projects in partnership with manufacturers of promising—but untested—technologies. In doing so, NREL could provide a range of technology-related services to assist in commercial deployment and speed access to capital, insurance products, and related market services.

As the industry grows quickly, the number of installers has grown quickly as well. NABCEP offers a voluntary certification process "by which PV installers with skills and experience can distinguish themselves from their competition. Certification provides a measure of protection to the public by giving them a credential for judging the competency of practitioners. It is not intended to prevent uncertified individuals from installing PV systems or to replace state licensure requirements" (NABCEP 2009).

NABCEP recently established PV installer certification and its newer solar thermal installer examination[14] in 22 locations throughout North America. These exams are intended to recognize significant experience and expertise of solar professionals.

In addition to its full practitioner certifications, NABCEP also offers an entry-level "Certificate of Knowledge" program for solar PV systems.[15] According to the NABCEP, 1,048 individuals have been awarded its Solar PV Installer Certification.

Several states require contractors to be NABCEP-certified for rebate approval (Coughlin 2009).[16] Extended reliance on the NABCEP certification process could increase quality and reduce accidents, thus leading to reduced insurance premiums.

In addition, opening communication among the insurance industry, the DOE, and the development community could lead to better-understood and improved guidelines, and thus lower costs of insuring PV systems.

The DOE might be able to lower the cost of insurance for renewable energy companies by providing limited protection from liability for relevant companies if it has this authority.

The U.S. Department of Homeland Security (Homeland Security) offers limited liability protection for "anti-terrorism technologies" via the SAFETY Act (Subtitle G of the Homeland Security Act of 2002),[17] which is designed to encourage the development of "anti-terrorism technologies" by providing manufacturers and sellers of these technologies with limited liability. The concern is many companies may not "invest" in these technologies without the SAFETY Act.

A similar liability protection could be offered to renewable energy companies (including developers, installers, and manufacturers) to promote investment and implementation in this area.

Or, Homeland Security could provide renewable energy projects with limited liability if the case could be made that PV is an anti-terrorism technology that qualifies under the SAFETY Act.

REFERENCES

Ace Limited. (2009). ACE USA. Ace Limited. http://www.acelimited.com/AceLimitedRoot/About+ACE/ACE+CompaniesInsurance+-+North+American/ACE+USA.htm. Accessed March 24, 2009.

AIU Holdings (AIG). (2009a). "Advanced Energy Solutions Group." American International Group. http://www.aig.com/_20_27295.html. Accessed March 24, 2009.

Chubb. (February 2008). "Insurance for the Renewable Energy Industry." http://www.chubb.com/businesses. Accessed March 24, 2009.

CompositesWorld.com. (June 1, 2008). "Suzlon Blade Recall: Retrofit Program to Address Cracking." June 1, 2009. http://www.compositesworld.com/news/suzlon-blade-recall-retrofit-program-to-address-cracking.aspx. Accessed November 30, 2009.

Cory, K.; Coggeshall, C.; Coughlin, J.; Kreycik, C. (2009). *Solar Photovoltaic Financing: Deployment by Federal Government Agencies.* NREL/TP-6A2-46397. Golden, CO: National Renewable Energy Laboratory.

EurObserv'ER. (2009). Photovoltaic Barometer. Paris, France: EurObserv'ER Consortium. http://www.eurobserv-er.org/downloads.asp. Accessed July 2009.

Greentech Media. (2009). "REC to Recall All of Its Solar Panels from 2008: Report." http://www.greentechmedia.com/green-light/. Accessed November 30, 2009.

Hartwig, R. P. (December 2008). "2008 - First Nine Months Results." New York, NY: Insurance Information Institute. http://www.iii.org/media. Accessed April 15, 2009.

Insurance Information Institute. (2009a). Glossary. Risk. http://www2.iii.org/glossary/R/. Accessed March 24, 2008.

Ins. Info. Inst. (2009b). Glossary. Law of Large Numbers. http://www2.iii.org/glossary/L/. Accessed November 25, 2008.

Ins. Info. Inst. (2009c). Contribution to GDP. http://www.iii.org/economics. Accessed March 23, 2009.
Ins. Info. Inst. (2009d). Facts and Statistics: Industry Overview, Insurance Industry at a Glance. http://www.iii.org/media. Accessed April 24, 2009.
Ins. Info. Inst. (2009e). Facts and Statistics: Insurance Company Rankings. http://www.iii.org/ media/facts/statsbyissue/insurancecompanyrankings/. Accessed April 15, 2009.
Ins. Info. Inst. (2009f). Glossary. Property/Casualty Insurance. http://www2.iii.org/glossary/P/. Accessed November 25, 2008.
International Energy Agency (IEA). (2008). *Trends in PV Applications: Survey Report of Selected IEA Countries between 1992 and 2007.* IEA-PVPS T1-17: 2008. Paris: International Energy Agency.
IREC. (2009). "NABCEP at 7: Its Positive Influence in Renewable Energy Workforce Development is Undeniable." http://www.irecusa.org/index.php?id=71&tx_ttnews[pS]= 1242201429&tx_ttnews[tt_news]=1428&tx_ttnews[backPid]=50&cHash =2532460b7d. Accessed June 8, 2009.
Jones, S. K. (2001). "Environmental Pollution Insurance: A Fluid and Ever-changing Market." *Insurance Journal.* http://www.insurancejournal.com/magazines features/18589.htm. Accessed April 24, 2009.
Manning, D. (2009). Personal communication. June 30, 2009. Hartford Financial Services Group.
Munich Re. (2005). *Topics Geo: Annual review: Natural catastrophes: 2005.* Munich, Germany: Munich Re. http://http://www.munichre.com/publications/302-04772 en.pdf. Accessed March 24, 2009.
North American Board of Certified Energy Practitioners (NABCEP). (2009). "PV Installer Certification." http://www.nabcep.org/certification. Accessed December 14, 2009.
National Association of Insurance Commissioners (NAIC). (March 17, 2009). "Insurance Regulators Adopt Climate Change Risk Disclosure." Press release. http://www.naic.org/ Releases/2009_docs/climate_change_risk_disclosure_adopted.htm. Accessed March 23, 2008.
Renewable Energy Policy Network for the 21st Century (REN21). (2009). "Renewables Global Status Report: 2009 Update." Paris: REN21.
Sherwood, L. (2008). "U.S. Solar Market Trends 2008." Latham, NY: Interstate Renewable Energy Council (IREC). http://irecusa.org/fileadmin/user_upload/NationalOutreachDocs/ SolarTrendsReports/IREC Solar Market Trends Report 2008.pdf. Accessed August 2009.

Signet Solar, Munich Re, and Marsh. (May 27, 2009). "Signet Solar, Munich Re and Marsh Cooperate on a New Type of Insurance Solution." Press release. http://www.signetsolar.com/pdf/ PISignetMunichRe270509Final_en.pdf. Accessed June 8, 2009.

U.S. Department of Energy (DOE). (December 2003). *A Consumer's Guide: Get Your Power from the Sun*. DOE/GO-102003-1844. Produced by the National Renewable Energy Laboratory for the DOE Office of Energy Efficiency and Renewable Energy (EERE). Washington, DC: EERE.

U.S. Department of the Treasury (Treasury). (2009). "Terms and Conditions: Payments for Specified Energy Property in Lieu of Tax Credits under the American Recovery and Reinvestment Act of 2009." http://www.treas.gov/recovery/docs/energy-terms-and-conditions.pdf. Accessed August 21, 2009.

Underwriters Laboratories (UL). (May 26, 2009). "Underwriters Laboratories Continues to Invest in its Capacity to Support the Global Growth of the Solar Industry." Press release. http://www.ul.com/global/eng/pages/corporate/newsroom/newsitem.jsp?n=underwriters-laboratories-continues-to-invest-in-its-capacity_20090526090800. Accessed June 8, 2009.

United National Environment Programme (UNEP). (2009). "Renewable Energy Insurance Training." Renewable Energy Insurance Training Kit - E-Learning Course by UNEP DTIE Energy Branch. http://www.energy. Accessed June 8, 2009.

Zurich Global Energy (Zurich). (2008). "Custom Risk Management Solutions: Zurich Alternative Energy." New York: Zurich Global Energy. http://www.zurichna.com/NR/rdonlyres/ 588FB711-0AA1-4A6D-8B3F-2657118D5471/0/ZurichAlternativeEnergy.pdf. Accessed March 24, 2009.

BIBLIOGRAPHY

Chubb. (August 25, 2007). "Chubb Assembles Team to Focus on Green Energy Insurance Solutions." Press release. http://www.chubb.com/corporate/chubb7645.html. Accessed April 27, 2009.

Grama S; Wayman, E.; Bradford, T. (2009). *Concentrating Solar Power - Technology, Costs, and Markets: A Guide to the Impact CSP Technologies will Have on the Solar and Broader Renewable Energy Markets through 2020: 2008 Industry Report*. Cambridge, MA: Prometheus Institute for Sustainable Development and Greentech Media.

Guice, J.; King, J. D. H. (February 14, 2008). "Solar Power Services: How PPAs are Changing the PV Value Chain." Greentech Media.

Mann, L. (2008). "IRS Reinterpretation Will Benefit Utilities' RE Investments." *North American Windpower* (September), pp. 72-73.

Marsh (2009). "Marsh Establishes Renewable Energy Team, Reinforcing Energy." Press release. http://www.marsh.co.uk/mediacentre/2007/pr20070522a.php. Accessed April 27, 2009.

Prometheus Institute for Sustainable Development and Greentech Media. (June 2008). "The Growth of Utility Scale PV." *PV News* (27:6); p. 3.

Solar Energy Industries Association (SEIA). (2009). "U.S. Solar Industry Year in Review, 2008. http://seia.org/galleries/pdf/2008_Year_in_Review-small.pdf. Accessed April 28, 2009.

Sun Edison. (2009). "Xcel Energy - Alamosa, CO - 8.22 MW: Project Profile." http://www.renewablenrg.com/projects/alamosa.php. Accessed April 28, 2009.

U.S. Congress. House. (2008). *Emergency Economic Stabilization Act of 2008.* H.R. 1424. 110th Cong., 2nd sess. http://www.govtrack.us/congress

APPENDIX A: INTERVIEWEES

The following contact information is for insurance brokers, underwriters, solar developers, and other organizations that agreed to be acknowledged for participating in interviews for this report.

Insurance Brokers

Edgewood Partners Insurance Center
John Greenfield, Principal 2000 Alameda de las Pulgas Suite 101
San Mateo, CA 94403 Tel: 650-295-4618
jgreenfield@edgewoodins.com
www.edgewoodins.com

Hub International Insurance Services
Charles A. Leone
400 Taylor Blvd. #300
Pleasant Hill, CA 94523
Tel: 925 609-6560
Tel: 415 272-7530 (m)
Charles.Leone@hubinternational.com

Marsh
Tel: 212-345-6000 global.marsh.com/
Insurance Underwriters **Chartis**
marineandenergy@chartisinsurance.com
www.chartisinsurance.com/us-advanced-energy
295 182636.html

Chubb Group
15 Mountain View Road
Warren, NJ 07059 Tel: 908-903-2000 Fax: 908-903-2027 Telex: 299719
www.chubb.com

GCube Insurance Services Inc.
3101 Westcoast Highway, Suite 100 Newport Beach, CA 92663
Tel: 760-880-1646 info@gcube-insurance.com
www.gcube-insurance.com

The Hartford
Drake Manning, Director
Strategic Design-Renewable Energy
One Hartford Plaza
Mailstop CO 1-46
Hartford, CT 06155
Tel: 860-547-9517
drake.manning@thehartford.com

The Hartford
James Gardiner, Assistant Vice President
Industry Markets
One Hartford Plaza
Mailstop CO 1-46
Hartford, CT 06155
Tel: 860-547-7994
james.gardiner@thehartford.com

Starr Technical Risk Agency, Inc.
Jim Devon, Vice President/Regional Manager
3353 Peachtree Road NE,

Suite 1000 Atlanta,
GA 30326
Tel: 404-946-1435
Tel: 770-313-8633 (m)
james.devon@cvstarrco.com
www.cvstarrco.com/cv/starrtech

Solar Developers
MMA Renewable Ventures
621 E. Pratt St., Suite 300
Baltimore, MD 21202
Tel: 443-263-2900
Fax: 410-727-5387
mmarenew.com

SunPower
Corporate Headquarters
3939 N. 1st Street San Jose,
CA 95134 Tel: 408-240-5500
Fax: 408-240-5400
us.sunpowercorp.com

Suntech Energy Solutions
71 Stevenson Street, 10th Floor San Francisco,
CA 94105 Tel: 866-966-6555 (toll free)
Tel: 415-882-9922
Fax: 415-882-9923
sales@suntechamerica.com
www.suntech-power.com

Tioga Energy
2755 Campus Drive, Suite 145 San Mateo, CA 94403
Tel: 877-333-9787 (toll free) Fax: 650-288-1011

Others
Fat Spaniel Technologies
2 W. Santa Clara Street, 5th Floor
San Jose, CA 95113-1824
Tel: 408-279-5262

Fax: 408-516-9111
www.fatspaniel.com
RMS
7015 Gateway Blvd. Newark,
CA 94560 Tel: 1-510-505-2500
Fax: 1-510-505-2501
info@rms.com
www.rms.com

APPENDIX B: INTERVIEW QUESTIONS

Liability/Property
1) From an installer's perspective, which insurance products are barriers for larger-scale solar PV systems/developments (e.g. property, liability)?
2) How often and to what degree do insurance premiums affect the feasibility of third-party installations?
3) Is there a threshold in system size beyond which insurance becomes a greater issue?
4) What insurance companies/brokers do you typically work with? Who do you prefer to work with and why?
5) Is it true that under a third-party power purchase agreement (PPA), the owner—not the host—has to get both property and liability insurance?
6) What are the primary reasons for—or drivers of—property insurance requirements (e.g., property/building policies, utility demands, interconnection standards)?
7) What are the main risks covered by a property policy (e.g., theft, business interruption, shipping)?
8) What are the primary reasons for—or drivers of—the liability insurance requirements? (e.g., utility demands, interconnection standards, property/building policies)?
9) What are the main risks covered by liability (e.g., worker's compensation, damage to other property/persons, environmental risk)?
10) Are there any differences with liability or property policies with the third-party lease model? For instance, maybe the third-party owner gets the property insurance and the host gets the liability insurance?

Other Risks
11) To what degree do guaranties against the following risks play into renewable energy projects?
    - Construction phase: Construction/completion Risk; counterparty risk
    - Operation phase: Performance risk; counterparty risk; Fuel supply/weather resource risk; credit risk
    - All phases: Workers' compensation; auto; financial risk; political risk; force majeure risk, e.g. hurricane, fire, wind

Costs/Price
12) Which costs more: the property insurance against damage or the liability insurance to protect utility workers?
13) Can you give a percentage breakdown for the two, as well as a percentage of total project costs that are needed annually?
14) How does this affect the end-user (consumer) price?
15) Do insurance companies have adequate information?

Adequate Information
16) Do you feel that insurance companies and/or brokers lack adequate information about solar PV systems? If so, what do they need help with?
    - Liability: Potential risks associated with system operation? Adequacy of interconnection standards?
    - Property insurance: Technology risks? Ability to withstand weather concerns?
    - Other risks?

Warranties
17) Do you or your manufacturers offer warranties? What is the typical duration?
18) What does the warranty cover (e.g. labor, equipment.)?
19) Do the warranties ever substitute or diminish the need for insurance? Do the warranties ever reduce premiums?

Installers
20) Do your installers have special certification for putting in place solar PV systems?
21) If so, does this help lower premiums for workers' compensation or any other insurance products?

Best Practices
22) Have particular states or utilities been easier or harder to work with in regards to insurance? Could you point to one that uses a best practice?

What can we do?
23) Are there any activities you would recommend the Department of Energy and the National Renewable Energy Laboratory undertake to address these barriers?
- Gathering of historical loss data?
- Technical information to dispel any myths about PV technology?
- Certification for installers?
- Warrantee guarantee backup by the DOE?
- Product testing for life expectancy? Wind resistance?
- Testing prototypes?
- Helping to create catastrophic loss models for PV

What can you do?
24) Can you aggregate data in an anonymous fashion?

Property
2. What are the primary drivers of property insurance requirements with solar PV systems (e.g., system owners, utility and state requirements, utility demands, interconnection standards, property/building policies)?
3. What is the property insurance protecting?

Liability
4. What are the primary sources of liability insurance requirements with solar PV systems (e.g., utility and state requirements, interconnection standards, system owners, property/building policies)?
5. What does liability insurance protect?

Other Risks
6. To what degree do guaranties against the following risks play into renewable energy projects?
- Construction phase: Construction/completion risk; counterparty risk
- Operation phase: Performance risk; counterparty risk; fuel supply/weather resource risk; credit risk

- All phases: Worker's compensation; auto; financial risk; political risk; force majeure risk, e.g. hurricane, fire, wind

Solar Industry
7. What is your view of the industry?
8. How do you define a prototype? Does this definition include a minimum number of hours of operation? Commercialization?
9. How do you feel about research and development? Testing?
10. What is the interaction between insurance and warranties? Do any warranties reduce insurance premiums?

Special Insurance Issues with Third-Party Ownership
11. Are you aware of any special issues—or advantages to—underwriting policies that use the third-party PPA ownership model?
12. Are there any additional issues with the third-party purchase price variance (PPV) *lease* model?

Competition
13. Who are your competitors in financing solar PV projects?
14. Are there recognizable industry leaders for this niche? Are they different for property insurance and liability insurance?

Adequate Information
15. Do you feel insurance companies have adequate information for understanding the insurance needs of solar PV technology?
16. How much experience do you have with PV technology? Do you have specific concerns?
17. Do you feel that others within your industry have appropriate knowledge about these technologies?
18. Are there challenges in understanding this technology? Like what? What information would help you?
19. What sources of historical loss data are used to assess the risk of solar energy installations?
20. Is information from Japan and Europe used?

Costs
21. What model or standards are used to establish the pricing, e.g. RMS,

NAICS codes? Property? Liability?
- Which costs more, the property insurance (i.e., against damage) or the liability insurance (i.e., to protect utility workers)?

22. Can you give a percentage breakdown between the two, as well as a percentage of total project costs that are needed annually?
23. How does this affect the end-user (consumer) price?

DOE's Role

24. Are there other issue areas where DOE could play a role (e.g. certification, testing)?

APPENDIX C: SOLAR PV TESTING AND CERTIFICATION LABORATORIES

The following is a non-exhaustive list of laboratories affiliated with International Laboratory Accreditation Cooperation (ILAC), an "international cooperation of laboratory and inspection accreditation bodies formed more than 30 years ago to help remove technical barriers to trade." See http://www.ilac.org/ for more information. The asterisks (*) indicate the four laboratories that conduct the performance-parameter testing on PV equipment that is required by the California Energy Commission.

Bodycote
2395 Speakman Drive
Mississauga, ON, Canada
L5K 1B3
Tel: 905 822-4111
Fax: 905 823-1446
www.bodycote.com

CIEMAT – PVlabDER
Avda. Complutense, 22, 28040
Madrid, Spain
Tel: 34-91-3466745
Fax: 34-91-3466037

Electronics Test & Development Centre
(IEC 61215 ONLY)
Ring Road, Peenya Industrial Estate Peenya,
Bangalore 56058, India

European Solar Test Installation
210 20 Ispra, VA, Italy
Tel: 39-0332-7869145
Fax: 39-0332-789268

Eurotest Laboratori SrL
Via dell'Industria, 18 35020 Brugine (PD)
* **Florida Solar Energy Center**
1679 Clearlake Road
Cocoa, FL 32922-5703
Tel: 321-638-1000
Fax: 321-638-1010

Fraunhofer ISE
Institut für Solare Energiesysteme
Heidenhofstr. 2
79110 Freiburg, Germany
Tel: +49 (0) 7 61/45 88-0
Fax: +49 (0) 7 61/45 88-90 00
www.ise.fraunhofer.de

Fundacion Cener—CIEMAT
(IEC 61215 ONLY) Avda.
Ciudad de la Innovación n°
7 31621- Sarriguren, Spain
Tel: 34-948-25-28-00
Fax: 34-948-27-07-74
 info@cener.com
* **Intertek Testing Services NA, Inc.**
(formerly known as ETL)

JET
5-14-12 Yoyogi, Shibuya-ku
Tokyo, 151-8545

Tel: 81-3-3466-5234
Fax: 81-3-3466-9219
Tokyo@jet.or.jp (technical)
info@jet.or.jp (others)

Metrology & Testing Center of China
Electronics Technology Group Corporation
No. 18th Research Institute

PI Photovoltaik Institut Berlin AG
Einsteinufer 25, D-10587 Berlin, Germany
Tel: 49-30-3142-5977 F
ax: 49-30-3142-6617
info@pi-berlin.com

* TÜV Rheinland Group
51105 Köln
Tel: 49-(0)221-806-0
Fax: +49 (0)221 806 114
internet@de.tuv.com

TÜV Rheinland Japan
Shin Yokohama Daini Center
Bldg. 3-19-5 Shin Yokohama,
Kohoku-ku Yokohama 222-0033
TÜV Rheinland PTL, LLC
(formerly known as ASU-PTL)
2210 South Roosevelt Street
Tempe, AZ 85282 Tel: 480-966-1700
info@tuvptl.com www.tuvptl.com

* UL Photovoltaic Technology Center of Excellence
2600 N.W. Lake Rd.
Camas, WA 98607-8542
Tel: 1-877-854-3577
Fax: 1.360.817.6278
cec.us@us.ul.com
www.ul.com

VDE
Merianstrasse 28
DE-63069 Offenbach
Tel: 49-69-8306-600
Fax: 49-69-8306-555
vde-institut@vde.com www.vde.com/en

---

[1] Residential owners who have ground-mounted solar PV systems installed may need additional insurance coverage. However, ground-mounted PV systems for residential customers are not common and are not within the scope of this report.

[2] Black line indicates trend.

[3] The National Association of Insurance Commissions (NAIC) should not be confused with the North American Industry Classification System (NAICS) codes that are discussed later in the report.

[4] Residential PV installations can usually be included under a homeowner's policy.

[5] Theft is only a risk for small rooftop PV systems after panels are installed, according to other insurers interviewed for this report.

[6] The federal investment tax credit is realized in the first year that operation begins but vests linearly over the first five years of a project at 20% of the 30% ITC each year (LBNL 2009). Thus, if a project is not rebuilt, the owner must repay the portion of the ITC not yet vested.

[7] At $5 per watt, this represents a 100 kW system.

[8] Federal tax incentives include the investment tax credit (ITC) and accelerated depreciation schedules (Modified Accelerated Cost Recovery System or MACRS).

[9] After year six, the ITC is no longer subject to the recapture rules applied by the IRS.

[10] The 2007, NAICS was described as including all of the U.S. Economic Classification Policy Committee recommendations for revision presented (FR 70 12390-12399). For more information, see http://www.ntis.gov/products/naics.aspx.

[11] Sarah Kurtz, National Renewable Energy Laboratory, personal communication, April–September 2009.

[12] The CEC requires testing for modules to be eligible for incentives offered by the California Solar Initiative, which are available to all market segments except for new home construction For more information, see http://www.californiasolarcenter.org/pdfs/utility/sb1/PV ELIGIBILITY PROCEDURE 20090605.pdf.

[13] Sarah Kurtz, National Renewable Energy Laboratory, personal communication, April–September 2009.

[14] For more information about the PV installer certification, see http://www.nabcep.org/certification/ pv-installer-certification. For more information about the solar thermal installer certification, see http://www.nabcep.org/certification/solar-thermal-installer-certification.

[15] For more information, see http://www.nabcep.org/certificates/entry-level-certificate

[16] Jason Coughlin, National Renewable Energy Laboratory, personal communication, September 2009.

[17] The SAFETY Act refers to the "Support Anti-terrorism by Fostering Effective Technologies Act."

INDEX

#

21st century, 106

A

access, ix, 4, 36, 37, 39, 44, 63, 76, 101, 102, 126, 128, 132
accessibility, 130
accounting, x, 8, 52, 66
accreditation, 144
adhesives, 64
advancements, x, 53
aesthetic, 59, 74, 75, 77
aesthetics, 55, 59, 74, 75, 76
agencies, 116, 123, 130
alternative energy, 130
American Recovery and Reinvestment Act, 135
American Recovery and Reinvestment Act of 2009, 135
appetite, 118, 125
architects, 60
assessment, 58, 86, 91, 93, 126
assets, 14, 23, 26, 44, 61, 78
authority, 133
aversion, 30

B

bankruptcies, 106
barriers, 18, 38, 60, 65, 66, 76, 104, 139, 141, 144
barriers to entry, 38
base, 7, 22, 65
benefits, 16, 18, 24, 27, 49, 64, 72, 118, 119
bonding, 116
bonds, 117
bottom-up, x, 52, 57, 61
breakdown, 140, 143
building code, 75
building-integrated PV (BIPV), ix, 52, 56
Business Interruption, 115
business model, xi, 7, 8, 38, 100, 127
businesses, 38, 101, 133
buyers, 95

C

capital outflow, 48
carbon, 1, 8, 17, 24, 27, 32, 36, 49
carbon dioxide, 1, 8
carbon emissions, 27
carbon policies, 17
case studies, 75

cash, viii, 2, 3, 12, 15, 24, 27, 45, 47, 61, 115, 121
cash flow, 15, 24, 27, 45, 47, 115, 121
catastrophes, 106, 134
CEC, 14, 40, 96, 130, 131, 147
certificate, 2, 147
certification, 75, 101, 110, 113, 128, 131, 132, 135, 141, 143, 147
challenges, xi, 12, 20, 38, 45, 47, 56, 60, 61, 65, 66, 69, 76, 100, 104, 113, 119, 125, 143
Chicago, 17
China, 60, 145
cities, 92
City, 17, 60, 78
classes, 6, 36
classification, 100, 127, 130
cleaning, 115
cleanup, 115
clients, 107
climate, 93, 106, 113, 135
climate change, 106, 107
climates, 72, 93
CO2, 1, 8, 17, 26, 49
coal, 49
coding, 101, 127
collaboration, 61
collateral, 112
combustion, 49
communication, xi, 99, 101, 133, 134
community, 101, 133
compensation, 101, 117, 127, 140, 141, 142
competition, x, 56, 99, 113, 116, 129, 132
competitive advantage, 45
competitiveness, 57
competitors, 143
complexity, 13, 18, 102
composites, 133
comprehension, 127, 128
configuration, 125
congress, 136
Congress, 136

consensus, 56
construction, 111, 114, 115, 116, 117, 122, 124, 130, 147
consumer price index, 1, 48
consumers, 54, 58, 74, 75
contamination, 115
conversations, 104
cooling, 72
cooperation, 144
coordination, 127
corrosion, 72
cost benefits, 63
covering, 102
CPI, 1, 48
crises, 105
criticism, 75
crystalline, x, 52, 53, 61, 62, 111, 130
crystalline silicon (c-Si), x, 52, 61
CT, 137, 138
customers, vii, viii, ix, xi, 2, 3, 4, 5, 6, 7, 9, 12, 13, 14, 16, 17, 19, 20, 22, 23, 24, 26, 27, 29, 30, 31, 32, 37, 38, 39, 43, 45, 100, 103, 107, 123, 128, 147

D

damages, 102, 122
data collection, xi, 99, 127
data set, 100
database, 100, 129
defects, 117
degradation, 16, 54, 55, 72, 73, 74, 96
degradation rate, 54, 55, 73, 74, 96
Department of Energy, 51, 59, 78, 79, 80, 81, 100, 104, 135, 141
Department of Homeland Security, 133
Department of Labor, 130
deployments, 102
depreciation, ix, 3, 4, 16, 17, 22, 24, 27, 28, 43, 45, 47, 118, 147
derivatives, 60
disclosure, 107, 113, 135
distribution, 128

Index

dominance, 123
draft, 105
durability, 65, 66, 75

E

earthquakes, 127
economic downturn, 105, 106
economic performance, vii, viii, ix, 2, 3, 4, 7, 8, 9, 14, 15, 18, 19, 21, 22, 23, 28, 30, 34, 35, 36, 38, 39, 45
economics, viii, 2, 3, 5, 6, 7, 8, 9, 13, 14, 15, 22, 24, 26, 37, 38, 48, 54, 118, 124, 134
economies of scale, x, 7, 49, 53, 120, 121
electricity, vii, viii, xi, 2, 3, 5, 6, 7, 8, 11, 13, 14, 15, 16, 17, 19, 20, 24, 25, 27, 28, 30, 32, 34, 37, 41, 42, 43, 44, 48, 49, 53, 56, 61, 71, 92, 94, 95, 96, 100, 116, 119
Emergency Economic Stabilization Act, 136
emission, 24
energy, viii, ix, xi, 1, 3, 13, 14, 16, 19, 22, 27, 29, 30, 44, 49, 51, 54, 57, 59, 71, 76, 78, 79, 94, 100, 101, 107, 109, 112, 116, 124, 125, 127, 128, 130, 135, 137, 143
energy efficiency, 29
engineering, 107
environment, 57, 73
environmental conditions, 72
environmental issues, 122
environments, 54, 72, 73, 76, 96, 132
equipment, vii, viii, 2, 6, 8, 44, 93, 95, 128, 131, 141, 144
equity, 9, 18, 27, 48, 118, 119, 123
EU, 49
Europe, 56, 60, 96, 109, 126, 131, 143
European market, 45
evidence, 96
evolution, 39
expertise, 107, 132

exposure, 48, 65, 107, 123, 127
extreme weather events, 124

F

families, 102
federal government, 100
Federal Government, 133
financial, 102, 104, 105, 106, 109, 113, 114, 117, 118, 120, 122, 132, 140, 142
financial community, 132
financial crisis, 109, 113
financial incentives, 118
financial support, 132
fixed rate, 119
flaws, 131
floods, 127
force, 75, 140, 142
formation, 107
framing, 13, 32, 38, 64, 65, 73, 93
France, 75, 95, 134
fuel prices, 33
funding, 131
funds, 116, 117, 132
future insurance premiums, xi, 100

G

gallium, 51
GDP, 1, 48, 105, 134
geography, 102
Georgia, 93
Germany, 109, 134, 145
goods and services, 105
grants, 132
Greece, 42
greenhouse, 106
greenhouse gas, 106
greenhouse gas emissions, 106
gross domestic product, 1, 48
growth, 59, 74, 76, 103, 109, 110
growth rate, 74, 110

guidelines, 75, 101, 133

H

Hawaii, 15, 93
health, 105
heat transfer, 73
history, x, 52, 102, 126, 128
Homeland Security Act, 133
homes, 27, 57, 58, 60, 93
host, xi, 58, 92, 95, 96, 100, 101, 119, 120, 123, 140
House, 136
humidity, 132
hurricanes, 106

I

ICC, 96
IEA, 51, 58, 60, 79, 92, 93, 94, 96, 111, 134
improvements, 17, 20, 27, 30, 38
income, 118, 123
India, 144
indium, 51
individuals, 132
industries, 9, 28, 101, 104, 113, 128, 130
industry, x, xi, 56, 59, 61, 62, 64, 66, 76, 99, 100, 101, 102, 104, 105, 106, 107, 109, 110, 111, 112, 113, 120, 122, 123, 125, 126, 127, 128, 129, 130, 132, 133, 134, 142, 143
infancy, 129
inflation, 48, 124
infrastructure, 125, 131
injury, 114, 115
insulation, 66
insurance policy, 107, 108, 122, 123
integration, 56, 57, 75, 76, 81, 95
integrators, 101
interest rates, 48
interface, 63

internal rate of return, vii, viii, 1, 2, 3, 4
Internal Revenue Service, 41
International Energy Agency, 51, 58, 79, 134
International Energy Agency (IEA), 58, 134
investment, vii, viii, ix, 1, 2, 3, 4, 6, 7, 9, 11, 12, 13, 14, 15, 19, 20, 21, 22, 25, 27, 28, 30, 31, 32, 36, 37, 38, 39, 45, 47, 48, 49, 107, 122, 133, 147
investments, 9, 11, 12, 13, 14, 17, 29, 30, 48
investors, 9, 15, 18, 45, 48, 102, 107, 108, 119, 123
issues, 45, 55, 57, 58, 74, 75, 77, 82, 120, 142, 143
Italy, 75, 78, 81, 95, 144

J

Japan, 60, 96, 109, 143, 146

L

landscape, 75
large-scale project developers, vii, 2
laws, 76
lead, ix, 3, 4, 9, 12, 14, 22, 27, 30, 33, 38, 43, 64, 101, 106, 118, 127, 128, 129, 131, 133
lending, 96
liability insurance, 114, 122, 140, 142, 143
life expectancy, 141
lifetime, 45, 71, 72, 96
light, 79, 134
limited liability, 133
loan guarantees, 132
lower prices, 22

Index

M

magazines, 134
majority, 37, 55, 96, 120, 121, 124
management, 113, 116, 129
manufacturing, 69, 76, 127, 128
market penetration, 39, 124
market segment, ix, 4, 6, 7, 9, 12, 18, 24, 36, 37, 38, 56, 147
market share, 60, 66, 75
marketplace, 56, 57
marsh, 136, 137
materials, ix, x, 52, 53, 56, 59, 60, 61, 63, 64, 65, 66, 68, 69, 75, 76, 81, 83, 84, 85, 86, 95, 131
matter, 75
media, 134
membership, 132
meter, 83, 84, 85
methodology, x, 52, 61, 63
Mexico, 96
military, 65
Missouri, 17
MMA, 138
models, 38, 39, 43, 66, 71, 127, 128, 141
modules, 53, 55, 56, 61, 62, 63, 64, 65, 72, 76, 79, 80, 81, 82, 92, 93, 95, 96, 115, 118, 125, 126, 127, 128, 130, 131, 147
multiple interpretations, 14

N

National Renewable Energy Laboratory, v, 1, 40, 41, 42, 51, 61, 78, 79, 80, 81, 99, 100, 133, 135, 141, 147
National Renewable Energy Laboratory (NREL), 41, 61, 100
natural disaster, 106, 113
natural disasters, 106, 113
natural evolution, 23
natural gas, 49, 125

net present value (NPV), vii, 2, 9
New England, 17
niche market, 125
North America, 101, 127, 128, 132, 135, 136, 147
NREL, x, 1, 8, 13, 14, 31, 32, 40, 41, 42, 43, 51, 52, 60, 63, 73, 78, 79, 80, 81, 86, 91, 92, 93, 96, 97, 99, 101, 104, 108, 118, 126, 127, 131, 133

O

officials, 38, 76
operating costs, 49, 118, 124
operating data, 129
operations, 2, 91, 119
opportunities, 14, 19, 32, 45, 49, 54, 56, 59, 61, 64, 65, 69, 74, 75, 76, 113, 126, 128, 129
ownership, vii, viii, 2, 6, 8, 13, 18, 21, 23, 27, 31, 37, 38, 43, 44, 47, 48, 49, 95, 96, 116, 118, 119, 120, 121, 123, 127, 143
ownership structure, 6, 8, 13, 18, 31, 43, 116

P

parity, 11
participants, vii, 2, 5, 15, 19, 23, 36, 38, 39, 105, 121
permission, 104, 110
personal communication, 147
photographs, 57
photovoltaic (PV), ix, x, 52, 55, 99
policy, ix, 4, 5, 6, 16, 17, 19, 23, 33, 36, 37, 39, 48, 49, 75, 76, 103, 104, 107, 115, 140, 147
policy makers, 104
policymakers, 39, 61
pollutants, 115
pollution, 115, 122

polypropylene, 65
poor performance, 116
portfolio, 119
potential benefits, 56, 65, 70
power generation, 114, 130
power plants, 125
preparation, 79
present value, vii, viii, 2, 3, 7, 9
preservation, 76
President, 56, 138
private sector, 131
probability, 88, 91
probability distribution, 88, 91
procurement, 116, 118
producers, 125
product design, 59, 60, 64, 66, 75, 77
professionals, 75, 100, 104, 127, 132
profit, vii, viii, 2, 6, 11, 24, 61, 64, 83
profitability, 2, 12
protection, 66, 107, 132, 133
prototype, 125, 126, 142
prototypes, 60, 96, 125, 141
public sector, 6, 12, 24, 26, 30, 38
public-private partnerships, 131
PVC, 80

Q

qualifications, 117
quality control, 117

R

radiation, 65
rate of return, viii, 3, 9, 17, 48, 49, 105
RE, 78, 136
recall, 133
recognition, 131
recommendations, 5, 147
recovery, 1, 131, 135
redevelopment, 115
reference system, 27, 34

regulations, 75, 106
relevance, 128
reliability, 65
renewable energy, x, xi, 2, 8, 37, 52, 100, 101, 104, 107, 108, 109, 112, 116, 118, 119, 125, 126, 127, 129, 130, 133, 140, 142
renewable energy technologies, 109, 125
repackaging, 32
repair, 115
requirements, xi, 63, 75, 77, 83, 84, 85, 96, 100, 105, 116, 122, 123, 132, 140, 142
researchers, 104
reserves, 27
Residential, v, 5, 10, 11, 14, 21, 40, 41, 42, 43, 51, 53, 61, 62, 63, 67, 68, 79, 82, 88, 91, 93, 103, 147
residential customers, vii, viii, 2, 6, 12, 14, 16, 27, 30, 32, 38, 43, 147
resistance, 141
resources, 91
response, 19
restoration, 115
restrictions, 132
retail, 5, 6, 7, 8, 11, 13, 15, 17, 19, 43, 44, 48, 66, 67, 83, 84, 85, 119
revenue, 11, 16, 30, 47, 105, 112
risk, xi, 9, 14, 17, 18, 48, 100, 101, 102, 104, 107, 109, 110, 112, 113, 114, 115, 116, 117, 123, 124, 125, 126, 127, 128, 129, 135, 140, 142, 143, 147
risk assessment, 123, 126, 127
risk management, 104, 113, 116, 117, 127
risks, 96, 100, 101, 102, 106, 107, 110, 111, 112, 114, 115, 116, 119, 121, 123, 125, 126, 127, 128, 130, 140, 141, 142
rules, 56, 147

S

safety, 65, 75, 116, 128, 130

Index

savings, viii, 2, 3, 5, 13, 14, 19, 22, 23, 32, 43, 44, 53
scale system, 6, 48
scaling, 19
science, 80
scope, viii, 3, 55, 64, 147
sellers, 95, 133
sensitivity, ix, 3, 5, 15, 17, 19, 20, 23, 24, 27, 29, 30, 34, 37, 73
service provider, 6
services, 58, 116, 117, 118, 129, 131
shape, vii, 2, 5, 15, 22, 38
shelter, 118
shingles, 53, 54, 62, 63, 65, 66, 67, 68, 81, 96
showing, 5, 19, 57
SIC, 101, 127, 130
silicon, x, 51, 52, 53, 60, 61, 62, 111, 130
simulation, 91
software, 127, 128
solar system, 54, 57, 62, 71, 72, 77
solution, 45, 47
Spain, 109, 144, 145
specialization, 107
stability, 65
stakeholders, 56, 60, 64, 74, 77, 100
standard deviation, 63
state, 7, 8, 16, 19, 38, 48, 49, 81, 95, 96, 101, 103, 123, 124, 132, 142
states, 8, 15, 17, 24, 36, 38, 93, 123, 132, 141
statistics, 129
structure, 6, 9, 44, 45, 48, 55, 72, 73, 74, 118, 119, 120, 121, 123, 130
Sun, 25, 40, 42, 78, 135, 136
suppliers, 59, 60, 116
supply chain, 58, 59, 86
supply curve, 58, 92
surface area, 62
Sustainable Development, 136
Switzerland, 76

T

target, 7, 69
tariff, 95
tax incentive, 12, 18, 116, 119, 147
tax rates, viii, 3, 16, 24, 28, 30, 44
taxation, 48
taxes, 8, 11, 45, 63, 64, 68, 82, 86, 95, 96, 118
techniques, xi, 62, 100, 128
technologies, x, 33, 52, 55, 60, 61, 64, 65, 66, 73, 94, 95, 100, 102, 104, 112, 113, 125, 126, 129, 131, 133
technology, x, xi, 18, 52, 61, 65, 66, 77, 93, 96, 100, 102, 104, 111, 112, 125, 126, 129, 131, 133, 141, 143
telephone, 104
temperature, 54, 55, 72, 73, 74, 96
terrorism, 133, 147
test data, xi, 100, 126
test procedure, 130
testing, xi, 72, 77, 99, 101, 113, 126, 127, 130, 131, 141, 143, 144, 147
theft, 114, 140
time series, 126
total energy, 8
total revenue, 45
trade, x, 52, 64, 144
trade-off, x, 52, 64
training, 128
transmission, 33, 52, 65, 128
transportation, 115
Treasury, 129, 132, 135

U

U.S. Department of the Treasury, 41, 135
U.S. economy, 105
UL, 127, 130, 131, 135, 146
underwriting, 104, 107, 108, 109, 112, 125, 129, 143

United, x, 40, 48, 58, 60, 61, 62, 67, 78, 79, 80, 86, 93, 94, 99, 103, 109, 113, 131, 135
United States, x, 40, 48, 58, 60, 61, 62, 67, 78, 79, 86, 93, 94, 99, 103, 109, 113, 131
up-front costs, 119
USA, 133

V

valuation, 19
Valuation, 42
vapor, 52
variables, 5, 11, 27, 30, 33, 44, 86, 91
variations, 8, 12, 27, 30, 32
vegetation, 93
Vice President, 138
vulnerability, 101

W

wage rate, 96
Washington, 41, 42, 135
water, 65
water vapor, 65
White House, 56, 81
wholesale, 6, 8, 11, 15, 19, 48
wind speed, 72
wind speeds, 72
Wisconsin, 80
workers, 101, 127, 140, 141, 143
worldwide, x, 37, 52, 60, 74

Y

yield, 93, 94